The Orce Man

Cultural Dynamics of Science

VOLUME 3

The titles published in this series are listed at *brill.com/cds*

The Orce Man

Controversy, Media and Politics in Human Origins Research

By

Miquel Carandell Baruzzi

BRILL

LEIDEN | BOSTON

This book received partial funding from the research project 'Museos, aulas y política: Cultura científica y tecnológica en la Transición española', funded by the Spanish Ministry of Science and Innovation [PID2019-104897GA-I00].

Cover Illustration: Replica of the VM-0 skull fragment as displayed in the Museo de Prehistoria, Orce. Photo by Nachosan, taken on 15 August 2015. The photo's background has been removed for use on the cover.

Library of Congress Cataloging-in-Publication Data

Names: Carandell Baruzzi, Miquel, 1981- author.
Title: The Orce man : Controversy, Media and Politics in Human Origins Research / By Miquel Carandell Baruzzi.
Description: Leiden ; Boston : BRILL, [2021] | Series: Cultural dynamics of science ; 2351-9932 Volume 3 | Includes bibliographical references and index. | Summary: "The Orce Man controversy is a very long, complicated and harsh dispute among scientific colleagues (often from the same institution) over the taxonomic classification of a cranial fragment. The fossil was found in 1982 by three palaeontologists from the Institut de Paleontologia de Sabadell in Catalonia: Josep Gibert, Jordi Agustí and Salvador Moyà-Solà at a site called Venta Micena, near the small town of Orce, located in a very arid region of the province of Granada, Andalusia"– Provided by publisher.
Identifiers: LCCN 2020034593 | ISBN 9789004431492 (hardback) | ISBN 9789004431508 (ebook)
Subjects: LCSH: Prehistoric peoples–Spain–Orce. | Human remains (Archaeology)–Spain–Orce. | Human beings–Origin. | Excavations (Archaeology)–Spain–Orce. | Venta Micena Site (Spain)
Classification: LCC GN836.O73 C37 2021 | DDC 936.6–dc23
LC record available at https://lccn.loc.gov/2020034593

Typeface for the Latin, Greek, and Cyrillic scripts: "Brill". See and download: brill.com/brill-typeface.

ISSN 2351-9932
ISBN 978-90-04-43149-2 (hardback)
ISBN 978-90-04-43150-8 (e-book)

This book is printed on acid-free paper and produced in a sustainable manner.

Al meu pare

Contents

Annexes

Acknowledgements

This book has only been possible thanks to the guidance and support of Oliver Hochadel. He introduced me to the Orce story and guided me in my first steps in history of science. The Institute for the History of Science at the *Universitat Autònoma de Barcelona* and its history of science master and doctoral program have been also essential to my formation as a historian of science. Formal discussions and informal talks with Agustí Nieto, Xavier Roqué, Alfons Zarzoso, Pepe Pardo, Annette Mülberger, Jon Arrizabalaga, Jorge Molero, Mònica Balltondre, Carlos Tabernero, Carlos Acosta, Lino Camprubí, Pep Simon, José Ramón Bertomeu, Pedro Ruíz, Emma Sallent, Sébastien Plutniak … and many others have completely influenced my way of thinking about history of science and shaped my career interests.

Xehc! student meetings with Joaquim Berenguer, Ferran Aragon, Judit Gil, Clara Florensa, Jaume Sastre, Jaume Valentines, Sara Fajula, Andrea Graus, Laura Valls, Òscar Montero, Gustavo Corral, Agustín López, Sergi Grau, Ned Somerville, Marcelo Valenzuela, Ignacio Suay, Mar Cuenca, Ignasi Medà, Xavier Ulled, Meritxell Ramírez and many others have been essential to ask everything you always wanted to know about history of science but you were afraid to ask. In my research stay in Cambridge, thanks to the British Society for the History of Science, Simon Schaffer was a unique sponsor and this book has a lot taken from talks with him. Cambridge had also the wonderful award of meeting Jesse Olszynko-Gryn and his family, Eoin Phillips, Peter Hellström, Julie Lawrence, Seb Falk and many others. Jesse and Eoin gave me great suggestions after reading and correcting fragments of this book. Suggestions by the two reviewers, by my editors Agustí, Oliver and Lissa, and by Veronika Lipphardt and Chris Manias were also very useful.

The images in this book are reproduced thanks to the permission given by the *Institut Català de Paleontologia de Sabadell*, the newspapers *Diari de Sabadell* and *La Opinión de Granada* and photographers Jordi Mestre, Sergio Ros-Montoya from the University of Málaga and Clara Florensa.

I have to thank Jordi Agustí, Lluís Gibert, Salvador Moyà-Solà, Domènec Campillo, Enrique García-Olivares, Michael Walker, Roger Marcet, Nora Moloney, Jorge Martínez Moreno, Eulàlia Subirà, Assumpció Malgosa, Àngel Galobart, Andrés Santos, Marc Furió, Agustí Camòs and Boris Santander for providing me with very informative interviews, informal chats or a great deal of unpublished documents that have enriched this book. Teresa Requena and Teresa Esquirol from the *Institut Català de Paleontologia* have been very helpful in the process to get into the Josep Gibert archive, which I prepared for archival deposit thanks to a Historical Archives Program Fellowship (HAP-125) of the Wenner-Gren Foundation. Translator Jean Byrne did a terrific job correcting my not very good English for most of this book, and Caroline Brew did it with the introduction.

Finally, I want to thank all my family and friends for their unconditional support. My brother, Tomàs, and his family, Carla, Paula and Mina, my inter-rail cousins and the rest of the Carandell, Baruzzi and Florensa families, too many to name all of them, and my friends are responsible for the outcome of this book. Clara, the Orce photographer, has always been the perfect partner in moral support and intellectual discussion in our (almost daily) hikes, t'estimo! My final word is for my mother Lory and her never-ending energy, thank you very much for everything.

Maps and Figures

Map

Figures

A Rough Guide to the Orce Man

As many actors, institutions, political parties, remains and sites were involved in the Orce Man story, I have put together this 'Rough Guide' with the aim of helping the reader to identify those names which repeatedly appear in the text. This 'guide' also provides the reader with a map to locate the sites, cities and towns relevant to the story (map 1).

Dramatis Personae
(*In alphabetical order*)

- *Emiliano Aguirre Enríquez* (*b.1925*): One of the best-known Spanish palaeoanthropologists of the second half of the 20th century and first director of the Atapuerca project. At the end of the controversy, he supported Gibert's views.

- *Jordi Agustí Ballester* (*b.1954*): Member of the Orce Man discovery team and director of the *Institut de Paleontologia de Sabadell*, after Gibert. He was later critical of the classification of the Orce Man. He was finally co-author of the paper which presented the Orce Boy.

- *Juan Luis Arsuaga* (*b.1954*): Atapuerca team co-director and one of the most widely known Spanish palaeoanthropologists (mainly due to his popular science books). He was highly critical of the Orce Man.

- *José María Bermúdez de Castro y Risueño* (*b. 1952*): Atapuerca team co-director and Orce conference attendee. He was critical of the Orce Man and was also co-author of the Orce Boy paper.

- *Domènec Campillo Valero* (*b.1927*): One of the first experts to see the Orce bone, claiming it to be a hominid. He never changed his opinion and continued to collaborate with Gibert.

- *Eudald Carbonell i Roura* (*b.1953*): 'First European' hunter, Atapuerca team co-director, and very critical of the Orce Man and Gibert's research. He is Catalonia's most famous archaeologist.

- *Yves Coppens* (*b.1934*): Well-known French anthropologist and one of Lucy's co-discoverers. He supported Gibert and was invited to the Orce conference but in the end did not attend.

- *Miquel Crusafont i Pairó* (*1910–1983*): Well-known palaeontologist during Franco's dictatorship when he founded and directed the *Institut*. He died at the beginning of our story.

- *Raymond Dart* (*1893–1988*): Australian scientist who first described *Australopithecus*. His story was used by Gibert as a mirror to his own story.

- *Enrique García-Olivares* (*b.?*): Spanish immunologist who performed immunological tests on the Orce bone. Since the first tests in the 1980s, he always supported Gibert's view that the remains were human.

- *Lluís Gibert Beotas* (*b.1968*): Son of Josep Gibert, he went to excavate in Orce at a very young age. Later he was a member of his father's team and then finally formed another team to try to continue excavations in Orce.

- *Josep Gibert i Clols* (*1941–2007*): The main character of the story. He was a member of the Orce Man discovery team, director of the *Institut* and the main defender of the Orce Man.

- *Donald Johanson* (*b.1943*): Famous American palaeoanthropologist and co-discoverer of Lucy. His books were widely read in Spain.

- *Richard Leakey* (*b.1944*): Famous Kenyan palaeoanthropologist and discoverer of well-known hominid remains. He was invited to the 1995 conference in Orce but did not attend.

- *Jerold Lowenstein* (*b.?*): American researcher who performed immunological tests on the Orce bone. He also attended the Orce conference and was a supporter of Gibert's views.

- *Henry de Lumley* (*b. 1934*): Prestigious French prehistorian. He wanted the discoverers to release an international publication on the Orce Man remains with him. Later he was very critical of the Orce Man. Husband of Marie-Antoinette.

- *Marie-Antoinette de Lumley* (*b. 1934*): Prestigious French anthropologist. She was the first to suggest the equine possibility to *El País*. Wife to Henry.

- *Bienvenido Martínez-Navarro* (*b.1964*): He went to Orce as a secondary school student and was a member of Gibert's team. He later changed sides and formed a new team. He was one of the discoverers of the Orce Boy in 2013.

- *Salvador Moyà-Solà* (*b.1955*): Member of the Orce Man discovery team. He was later critical of the Orce Man classification and was director of the *Institut* after Agustí.
- *Paul Palmqvist Barrena* (*b.?*): Member of Gibert's team. He was later critical of the Orce Man and of Gibert and was one of the discoverers of the Orce Boy.
- *Joan Pons Moyà* (*1955–2003*): *Institut* collaborator. He was the discoverer of Cueva Victoria and of the phalanx found there attributed to a hominid.
- *Pascual Rivas Carrera* (*b.1945*): One of the first members of the Orce research team. He had political relevance and was later critical of Gibert. He became head of the Orce research *International Commission.*
- *Phillip V. Tobias* (*1925–2012*): Famous South African palaeoanthropologist. He supported Gibert's views during and after the 1995 Orce conference.
- *Isidro Toro Moyano* (*b.?*): One of the first members of the Orce research team. He had political relevance and was later critical of Gibert. He was also one of the discoverers of the Orce Boy.
- *Jaume Truyols Santoja* (*1921–2013*): Leading palaeontologist. He first praised but later criticised Gibert and his handling of the Orce discovery.
- *Alain Turq* (*b.?*): French prehistorian who first excavated at Fuente Nueva 3. He supported Gibert but later changed sides publicly.
- *Michael J. Walker* (*b.1941*): British archaeologist who collaborated with Gibert and directed the excavations at the Cabezo Gordo Neanderthal sites.

Institutions

- *Diputación Provincial de Barcelona/Diputació de Barcelona*: Barcelona's provincial council, which after Franco's dictatorship no longer had powers nor a clear aim. It was very active in publicising the Orce Man discovery in 1983, when it was controlled by the PSC-PSOE.
- *Diputación Provincial de Granada*: Granada's provincial council, it was first excluded from the management of the Orce Man discovery.

- *Generalitat de Catalunya*: The Catalan government, established after political autonomy was granted in 1979. Controlled by the right-wing Catalan nationalist party, *Convergència i Unió* (CiU). It did not participate in the public presentation of the discovery of Orce Man.

- *Junta de Andalucía*: The Andalusian government established after political autonomy was granted in 1981 and controlled by the *Partido Socialista Obrero Español* (PSOE) until 2019. In 1983, it actively participated in the public dissemination of the Orce Man.

- *Institut de Paleontologia de Sabadell*: The scientific institution where the Orce Man discoverers worked and gained permanent positions after the discovery. Since 1983, the three discoverers have each been director of the institute. Politically, the *Institut* was overseen by the *Diputació de Barcelona.*

- *Museu d'Arqueologia de Catalunya*: Also under the *Diputació de Barcelona*, it was the place where the inner part of the Orce Man bone was cleaned. It was also Domènec Campillo's home institution.

- *Museo Arqueológico de Granada*: Main archaeological institution in Granada and the place where archaeologist Isidro Toro worked for many years.

- *Universidad de Granada*: Granada's university, where immunologist Enrique García-Olivares and palaeontologist Pascual Rivas worked for many years.

- *Museo de Prehistoria y Paleontología de Orce, José Gibert*: Orce's local museum, new home to the Orce Man and the rest of the Venta Micena remains. It was the institution where Bienvenido Martínez-Navarro worked until he was dismissed in 1996. A new museum was inaugurated in August 2015.

Political Parties

- PSOE – *Partido Socialista Obrero Español*: Spanish socialist party which came to power in the 1982 general elections and dominated the Andalusian government.

- IU – *Izquierda Unida*: Left-wing Spanish party which supported the right-wing PP for some years in Andalusia, allowing Gibert to get funding for the 1995 conference in Orce.

- *PP – Partido Popular (formerly AP – Alianza Popular)*: Right-wing Spanish party which supported Gibert during the conference and after.

- *PSC-PSOE – Partit dels Socialistes de Catalunya*: Catalan PSOE, which dominated several cities around the Barcelona province and therefore controlled the *Diputació de Barcelona.*

- *PSUC – Partit Socialista Unificat de Catalunya*: Left-wing Catalan party of which Gibert was a member.

- *PCC – Partit dels Comunistes de Catalunya*: Catalan pro-Soviet party which emerged from the PSUC and for which Gibert presented in the 1986 general elections as number 10 on the electoral list.

Remains

A brief note to highlight that, despite the fact that the current usage in palaeoanthropology today has changed (not without dispute) from 'hominid' to 'hominin' to refer to members of the human evolutionary line, I have preferred to still use 'hominid' as it was the word used by the historical actors. This way, my historical narrative of the Orce story fits better with the primary sources used.

- *VM-0*: The famous Orce Man, this story's 'bone of contention'. A less than 10 cm cranial fragment with parts of the parietal and occipital bones. For more anatomical details, see Annex 1.

- *VM-1960 and VM-3691*: Two humerus fragments found by Gibert's team in Venta Micena and claimed to be from hominids. Much less known, even for specialists, than VM-0.

- *BL-0*: A supposed hominid tooth fragment found at Barranco León, presented during the 1995 conference in Orce and later published by Gibert's team and ignored by the discoverers of the 'Orce Boy'.

- *BL02-J54-100*: A hominid milk tooth named the 'Orce Boy'. Found in 2002 in Barranco León and presented in 2013 by Isidro Toro, Bienvenido Martínez-Navarro and others as the earliest human remains in Europe.

- *CV-0*: A supposed hominid phalanx from Cueva Victoria found by Joan Pons-Moyà in 1984.

Sites

- *Venta Micena*: Discovered in 1976 in Orce. The site where the remains of the famous Orce Man (VM-0) were found. Later closed for several years and recently excavated but considered an only ancient fauna site.

- *Barranco León*: Discovered in 1987 by Gibert and collaborators in the Orce area. A site with widely recognised stone tools, a controversial fragment of a hominid tooth (BL-0), and the place where the Orce Boy (BL02-J54-100) was discovered.

- *Fuente Nueva 3*: Discovered in 1992 in the Orce area. A site with widely recognised stone tools, first excavated by the French prehistorian Alain Turq.

- *Cueva Victoria*: A cave in Cartagena discovered by Pons Moyà, where a supposed hominid phalanx was found in 1984. Also excavated by Gibert's team over a long period of time.

- *Cabezo Gordo*: Neanderthal sites excavated by Michael J. Walker in collaboration with Gibert. The research carried out there was presented at the 1995 conference in Orce.

- *Almenara-Casablanca*: An area where stone tools of a purported 'First European' were found and presented to the public in 1983. Not much later, these stone tools turned out not to be hominid-made.

- *El Aculadero*: A site near Cadiz that was also claimed to have traces of the 'First Europeans' in 1983. The evidence later turned out to be more recent.

- *Atapuerca*: A large complex of palaeoanthropological sites where thousands of hominid fossils and stone tools have been found. Thanks to a great popularisation effort, the site has become a household name in Spain.

- *Altamira*: A Spanish cave with famous rock paintings claimed to be prehistoric, some of the first to be discovered (1879). Claimed to have caused a controversy between Spanish and French scholars that ended with a 'Mea culpa d'un sceptique'.

MAP 1 Map of Spain with cities and sites relevant to the story

Introduction—The Orce Man

The Orce Man controversy is a very long, complicated and harsh dispute among scientific colleagues (often from the same institution) over the taxonomic classification of a cranial fragment (figure 1). The fossil was found in 1982 by three palaeontologists from the *Institut de Paleontologia de Sabadell* in Catalonia: Josep Gibert, Jordi Agustí and Salvador Moyà-Solà at a site called Venta Micena, near the small town of Orce, located in a very arid region of the province of Granada, Andalusia (map 1).

The three researchers soon concluded that the cranial fragment belonged to the oldest ancestor of humankind in Europe. Yet, only the external part of the fragment was visible, as the inner part was covered by a piece of rock that was impossible to extract without proper equipment. At that point, they contacted the politicians in charge of their scientific institution to show them this new and important find. A scientific publication on the fragment was arranged together with a press conference to present the discovery to the Spanish public in 1983.

The Orce Man attracted a great media following with coverage from several newspapers, magazines and television channels.[1] The political situation in Spain, with democracy just emerging after 40 years of dictatorship, was a crucial factor in the way that this discovery was disseminated. After a year, during which the discoverers appeared in the media quite often, the internal part of the fragment was cleaned and a strange crest appeared.

This crest seemed to indicate that the fragment was not from a hominid but rather belonged to an ancient equine.[2] The new information was first made public on the front page of the *El País* newspaper without the agreement of the discoverers. With this, a very public controversy broke out. Of the three discoverers, only Josep Gibert stuck to his position, maintaining that the Orce fragment belonged indeed to a hominid. During this period, almost the only space for scientific debate was the media. Three years later, in 1987, Moyà-Solà and Agustí announced to the media that they had changed their opinion and now agreed with the equine classification. Soon after, they published the first scientific article on this new classification.

In 1995, after years of scientific study on the cranial fragment and other remains, Gibert managed to organise an international conference in Orce,

1 For more on media attention to Orce see Annex 2.

2 For more anatomical details see Annex 1.

 | DOI:10.1163/9789004431508_002

FIGURE 1 VM-0 as presented at the *Orce Museum* in 2015
PHOTO CLARA FLORENSA

followed closely by the media, at which his claims (and therefore his reputation) were supported by several well-known palaeoanthropologists from abroad as well as politicians who promised funding and permits. Sometime after the conference, some of his younger collaborators again appeared in the media to state their disagreement with Gibert's claims. This new dispute led to the withdrawal of the politicians' pledges. From then on, Gibert was gradually excluded from the Orce research project until he died in 2007.

In 2013, some of Gibert's former colleagues published a scientific article in which they presented a tooth which they claimed to be the oldest human fossil in Europe. When this article was published on the journal's website, it did not include a single reference to Gibert or his work. Some supporters of Gibert's ideas, including his son Lluís, sent letters to the editor complaining about this omission. The article was then withdrawn and a small yet significant online dispute took place until the article was republished with references to Gibert's work.

Due to its strong public character and the accusations made between parties, the Orce story is often presented as an exception; a rarity in Spanish history of science that is unlikely to be repeated. Yet, while this book will highlight the complexities of the Orce case, it will also try to show how many of these supposed particularities are typical of palaeoanthropological research or even intrinsically part of scientific practice. In order to better understand how the Orce Man story tells a lot about scientific, and particularly palaeoanthropological research, a comparative case is also included. This case deals with a possible 'First' American in Brazil in a very similar period to the Orce Man.

Even a brief account of this case demonstrates the fragility of claims to the 'uniqueness' of the Orce Man and shows how its analysis could be applied in many other instances.

∴

To analyse the Orce controversy, this book draws a chronological narrative of the controversy, respecting as much as possible the sequence of events.[3] This allows us to analyse these events in their particular context and thus reduces interpretations of the facts based on later developments in the story. At the same time, this book uses the methodological approach of symmetry put forward by sociologist David Bloor.[4] According to this approach, the historian or sociologist must apply the same explanations to those scientific claims that are taken to be true and those that are taken to be false, since both are shaped and constructed by the same social processes. Thus unlike most, if not all, of the accounts of the Orce controversy, this book does not have as its aim to explain who was right and why. It is not important for us whether the Orce bone belonged to a hominid or not. Instead, this book will attempt to understand all of the different actors' positions in an effort to treat both sides of the controversy equally. What interests us is the examination of the strategies and circumstances that make a certain claim credible or not for both the scientific community and the general public. This book is more concerned with how and why a certain claim is credible rather than how and why it is true.

Each of the four main parts of the book (discovery, controversy, conference and end) start with a brief 'opening vignette' (as one of the reviewers of the book manuscript put it). These are brief narrative scenes that, despite being based on reality, are fictional scenes. They are useful to situate the reader in the action and to introduce the chapter and its crucial moments. At the same time, these 'opening vignettes' are short exercises that, as Kathryn M. Olesko highlights in her reflections on the relationship between narrative structure and historical evidence, try to capture "multiple points of view or truths that literary writing is capable of exploring and expressing", while also deal with "contradictions and incompatibilities in the way historical writing cannot".[5]

The Orce Man story has been told several times, mostly by the historical actors themselves, the most noteworthy example being Gibert's 2004 popular

3 For more on narratives in history of science, see: Golinski 1998, 186–206.

4 For instance, Lynch 1994, 76–80.

5 Olesko 2007.

science book.[6] Yet, this story has rarely been narrated by actors outside of the controversy itself. Besides my own work, only the rhetoric of the early controversy and its relation with another important site in Spain, Atapuerca, has raised interest in the history or sociology of science community.[7]

To tackle this controversy, one of the main sources that I used was newspapers, complemented by magazines and other publications for the general public.[8] Six national newspapers composed the primary set of newspapers. They represent the different ideological, geographical and historical positions of the country.[9] Hundreds of scientific articles have been published on the Orce site and the cranial fragment, by both sides of the controversy. There are approximately 15 to 20 papers that have been relevant to the story and its unfolding and therefore have been analysed in depth.[10] Similarly, my research has benefited from the consultation of personal archives of some of the main characters in the story. Video and audio recordings have also been very important for my research on Orce. At least three documentaries deal exclusively or partially with the Orce Man controversy. Yet, mostly, they present the controversy from Gibert's point of view.[11] I also conducted interviews with some of the main characters of the story.[12] Besides, a recorded interview with Josep Gibert before he died was also used. I am fully aware of the benefits and drawbacks that this kind of source oral history can entail for historical work.[13] Thus, although these

6 Gibert 2004. Other examples: Martínez-Navarro 1993; Gibert 1999a; Campillo 2002; and Iglesias/Gibert 2003. And in relation with the media Gibert/Gibert 2000.

7 Carandell 2012; Carandell 2013a; Carandell 2013b; Carandell 2015a; Carandell 2015b; Díaz Rojo 2007a; Díaz Rojo 2007b; Díaz Rojo, 2011; and Moreno 2014.

8 All of the literature used throughout this work is gathered in the bibliography. In the footnotes, a reduced version of the references will allow the reader to find the complete reference in the bibliography. This bibliography has been divided into four parts: interviews (I); archives (II); secondary literature, including scientific literature, popular books, webpages, and video recordings (III); and a final section devoted exclusively to newspaper pieces (IV). This last part will be easily recognised in the footnotes, as the reference will include the name of the newspaper in italics, as opposed to just the year, page number, etc. For instance: '*González, El Periódico, ABC* 1983', the italics *ABC* means that to find the complete version, it is necessary to check the newspaper section of the bibliography.

9 Two traditional and conservative newspapers both dating back to the 19th century: *ABC*, published in Madrid and Seville, and *La Vanguardia*, published in Barcelona; two newspapers that represent the left-wing ideas of the new democratic state, *El País*, published in Madrid, and *El Periódico*, published in Barcelona; and two local papers from Sabadell, *Diari de Sabadell*, and Granada, *El Ideal de Granada*.

10 For the complete references for these articles see the bibliography.

11 Guàrdia/Pou 1996; Balart 2007; and Navarro 2010.

12 See a list of the interviews in the bibliography.

13 For more on the use of the oral history of science, see for instance: Chadarevian 1997.

interviews have made up one of the main sources for this book, together with newspaper articles, I use them only to supplement information that I have gained through other sources or to fill gaps in the story that do not appear in written accounts of it. In all of these cases, I exercise caution and always underline that the information comes from oral interviews made much more recently than many of the events being discussed.

All of these sources have allowed a close analysis of the Orce Man story, at some points following its development almost day to day. This analysis has shown the relevance of this story in the study of three aspects of scientific practice that almost constantly overlap: politics, controversies and media. In what follows, I will present how this book analyses these three aspects and how they affect the validation and credibility of a specific scientific knowledge claim. I will show how the Orce Man story fits into the work done by historians of science on controversies, media, credibility, boundaries of science and the history of palaeoanthropology, in order to highlight the contributions of this book to these historiographies.

∴

The study of scientific public and political controversies in the historical earth sciences has been of great importance in both the history of science and the sociology of science. In this regard, the present work's close examination of the Orce Man controversy aims to contribute to these studies by showing how this late 20th century public and political controversy developed in a particular social, scientific and political context. The detailed description of the find and the public presentation of a scientific discovery in this particular political context is another of the original contributions of this book. This spectacular find later turned into a long and harsh controversy. The detailed analysis of this scientific story of a failure may also serve as an interesting contribution to the history of palaeoanthropology, as history of science often forgets about failures and highlights success. The way in which I analyse the strategies used by different actors during the controversy to support claims and gain credibility, especially in the specific context of an international conference, sheds some new light on the characteristics of scientific disputes in these sciences and also on how meetings were used in this context. And finally, I hope to tell a story, a poignant and intense, scientific story of the Orce man.

One of the main aims of this book is to analyse the relations between scientific practices, popularization and politics. Specifically, it will explore how science had a significant role in the process of 'state-building' which Spain was experiencing in the early 1980s, after forty years of General Francisco Franco's dictatorship. At the same time, it highlights the role of the media generally, with the Orce Man case as an example, in the construction of science in this new democratic state.

During this period, Spain changed radically: political parties, public demonstrations, unions, contraception and divorce, among many other things, were legalised, and people could vote for the first time after the dictatorship. The Spanish state also changed its territorial organisation. Spain was divided into *autonomías* (autonomous communities) with their own regional governments. Due to historical and cultural demands, Catalonia was, together with the Basque Country, one of the first *autonomías* to get its own government, followed by Andalusia. Thus, by the summer of 1982, when the Orce Man was found, both territories, Catalonia (where the researchers were from) and Andalusia (where the remains were found), had their own recently created democratic parliaments (see map 1). This early democratic state, therefore, represented a period of opportunities for the younger generation of politicians who begun to occupy significant positions.[14] On 28 October 1982, the *Partido Socialista Obrero Español* (PSOE), the Spanish Socialist Workers' Party, won the general elections. The PSOE was the first left-wing party in government since 1936; it was the first time that the 'losing side' (*los vencidos*) of the Spanish Civil War got back into power. For this and other reasons, that day is considered to mark a change, to be a key moment in the Transition period and therefore the consolidation of the democratic state.[15]

For the media, the Transition and this early democratic consolidation represented a new free and open stage where people and journalists expressed themselves with almost no censorship or fear of retaliation. Therefore, these were times of political, social, cultural and also scientific, euphoria, after 40 years of a repressive dictatorship. This period has been characterised as showing a "vitalist and euphoric reaction" that generated new cultural and social spaces, "with an open, receptive and critical discourse" that did not have its place in politics but in the street, where social movements were crucial factors to political changes.[16]

This context also engendered in the media a particular willingness to report on scientific achievements, and especially on Spanish scientific achievements.[17] Popularization of science as such in Spain in the early 1980s was

14 González, *El Periódico* 1983.
15 For a general account on the Spanish Transition, see Molinero/Ysàs 2018.
16 Quotes García Naharro 2011, p.443, for the role of social movements, see Domènech 2002. For more on Catalan newspapers during this period, Cardús 1999.
17 Muniesa 2005, 173–176; González-Silva 2009; and González-Silva 2014.

practically non-existent. For instance, the use of popular science books by scientists or journalists – one of the main media used by paleoanthropologists in the international context – had been largely absent in Spain. The Orce Man was a first failed attempt to publish these books that would later become much more widely used by other Spanish palaeoanthropologists.

Therefore, the story of the Orce Man provides an important example of how science and its popularization were formed within the process of state-building. Scientific structures in many senses were built with state structures. The changes that took place during this period in Spanish history were vast and science was no exception. To explore how these changes occurred through the Orce story is one of the aims of this book. This role of science in the 'building of democracy' is also relevant because historical actors were aware of the process that their society was undergoing, and therefore it conditioned their actions and decisions, and thus conditioned scientific development. The Orce Man story is, therefore, a contribution to better understanding this period of intense transformation, and to understanding how scientists, journalists and politicians acted. Science had to be a crucial part of the new democratic country and therefore had to be spectacular, with big headlines and public presentations. Therefore, the Orce Man shows us how science had to be popularised in this new state.

The early 1990s was another period of intense euphoria in Spanish society.[18] During this period, European palaeoanthropology was experiencing what was called the 'hunt for the First European': a race to find traces of the oldest human ancestor on the continent and the attempt to push back the chronology of its arrival. A 'hunt' in which the Orce Man case is framed. This scientific context has so far received very little attention from the history of science, the Atapuerca site (also in Spain) being the only exception.[19] Atapuerca somehow symbolises both Spanish society's euphoria and the 'hunt for the First European'. In 1978, Emiliano Aguirre, one of the best-known Spanish paleoanthropologists of the second half of the 20th century, began an excavation at different sites in the Atapuerca Mountains, near Burgos (see map 1). During its first excavation years, Atapuerca was a promising site at which young Spanish researchers wanted to work. In 1991, Aguirre retired and the direction of the excavation passed on to three of his disciples: Juan Luis Arsuaga, José María Bermúdez de Castro and Eudald Carbonell. The trio began an intensification of the excavation work and hominid fossils appeared.

---

18 Economic growth, the Barcelona Olympic Games, the Seville Expo 92, or Indurain's victories in the Tour de France were some of the reasons for this euphoria. Muniesa 2005, 223.

19 Hochadel 2013b.

From then on, the research team produced spectacular discoveries year after year during the 1990s. Just a brief outline of their most important achievements conveys the scale and rate of their finds: in 1992, some complete craniums of *Homo heidelbergensis* were found together with thousands of other remains from these hominids; in 1994, what were considered the oldest European hominid remains at the time (around 780,000 years old) were found; in 1997, and for the first time in Spanish science, the Atapuerca directors named a new hominid species (*Homo antecessor*) on the basis of these remains; and, in 2007, a new hominid jawbone was unearthed and again labelled as the oldest European, now at 1.2 million years old. Atapuerca became one of the biggest projects in Spanish science, definitely the biggest in palaeoanthropological, palaeontological and archaeological science, and was often presented as an example to be followed. Thanks to Atapuerca, Spain is the second country, after the United States, with the most scientific publications on human evolution. In 1997, the Atapuerca project was awarded the *Premio Príncipe de Asturias*, the top Spanish prize, a sort of Spanish Nobel and the Atapuerca sites were declared a *UNESCO* World Heritage Site in 2000. Together with all of these discoveries, the Atapuerca team has engaged in very fruitful popularisation efforts that include several books written for the general public, documentaries and the spectacular *Museum of Human Evolution* in Burgos. The three Atapuerca co-directors have become well-known personalities among the Spanish general public. Moreover, apparently, relations between the Atapuerca researchers and politicians have been very close, resulting in steady public funding for the site's research and popularisation.[20]

During the same time, the Orce controversy also experienced a high point in its story. In 1995, an international conference was organized and the Orce discoveries went public again. The political context was crucial to allow the realization and success of the conference. The conference funding and the Orce research was debated in the Spanish and Andalusian parliaments, making this story an instructive example of the relation of scientific controversies with apparently well-established democracies. Seemingly very favourable to Gibert, the final result of the conference actually turned sour for him when the opinion of some researchers on the cranial fragment of some appeared in newspapers and on television. The Atapuerca researchers were crucial in this change in the public perception of Gibert's research.

20 For a detailed account of the Atapuerca story, see Hochadel 2013b, and for the uses of popular science books: Hochadel 2013c.

Atapuerca is a vitally important and very public story of Spanish scientific success; a story that has several points of contact with the Orce Man case. Yet unlike Atapuerca, the Orce story is not one of success but of failure; a scientific controversy that blocked research for years. Everybody in Spain knows Atapuerca, while Orce is known mainly by those interested in the topic or by those who remember the past controversies.[21] The politics behind both success and failure are crucial to understanding these processes. This contrast between both stories may help to answer several questions: What are the differences between success and failure in science? Which is the role of politics behind them? Do these differences depend only on the importance and quantity of the findings? Is the public presentation of the findings different depending on the political support for the claims? Do scientific colleagues react similarly to this popularisation? The analysis of this scientific failure, how it happened and how it was maintained over time, is one of the aims of this book.

∴

Palaeoanthropology is a very controversy-prone science. Human fossils are difficult to find and when discovered they are often very fragmented and supply only limited information. It is very unusual to discover an almost complete skeleton or a complete skull and often palaeoanthropologists work for years at sites and discover very few remains. But there is much more to palaeoanthropological science than that. Animal fossils, stone tools, dating methods and several other techniques provide a lot of information about our ancestors. Yet, the attention of the general public and the funding bodies is still captured by the few and hard-won hominid remains. Because of this, when something is found, individual researchers tend to emphasise the significance of their own discoveries for human genealogy in order to increase the audience's favour and therefore, to get more funding. This can collide with other researchers' interests and often leads to open rivalries and debates. Due to the public character of palaeoanthropology, as the Orce Man case shows, what starts out as a restricted scientific skirmish can turn into a very public controversy.[22]

Therefore, the history of palaeoanthropology and prehistory is full of long and arduous controversies that are an integral part of it and have shaped its development. Some of these stories, and also their main characters, will appear repeatedly in the Orce story and therefore, form a crucial part of its historical

21 Hochadel 2013b, 26–27.

22 Several examples in Lewin 1997.

and scientific context. Below, these stories will be briefly introduced in order to better understand these contexts and how they influenced the Orce story.

The most cited palaeo-controversy story involving the Orce actors, probably because it is a Spanish one, was the Altamira cave discovery controversy (see map 1). The traditional 'official' story of Altamira starts when the amateur archaeologist Marcelino Sanz de Sautuola was looking for prehistoric artefacts and found the Altamira rock paintings in 1879.[23] Since some animals represented were extinct in Europe, Sautuola deduced that the images were from prehistoric times. Together with the Spanish prehistorian Juan Vilanova y Piera, he presented the discovery to the international scientific community at several congresses. Subsequently, the prestigious French archaeological journal *Matériaux pour l'histoire primitive et naturelle de l'homme* asked Édouard Harlé to write a report on the paintings. After examining them, Harlé concluded that they had been made recently. His verdict influenced the rest of French prehistorians and archaeologists, including Émile Cartailhac, one of the most prominent archaeologists at the time, who denied the validity of Altamira's cave paintings and to some extent ridiculed Sanz de Sautuola in his public presentations. This 'official' story continued when the later discovery of similar paintings in French caves such as Font-de-Gaume led the French 'academic establishment' to accept the Spanish prehistoric rock paintings. In 1902, Cartailhac wrote a famous article, 'Mea culpa d'un sceptique', in which he admitted the authenticity of Altamira.

Another frequently cited palaeo-controversy was that of the Taung Boy. In 1924, Raymond Dart, an Australian anatomist, settled in South Africa, received a box full of fossils from the Taung mine.[24] Among them, there was a cranium with a human appearance. Dart published an article on the cranium in the journal *Nature* stating that it was an extinct "race of apes intermediate between living anthropoids and man" that he named *Australopithecus*.[25] Apparently, this statement was enough for the British scientific authorities to turn against Dart's discovery. The controversy lasted for more than 25 years and only ended with new discoveries and another English authority (Wilfred Le Gros Clark) supporting Dart's views.

23 For the 'official' story of Altamira, see Calvo 2015 and Madariaga 2000. For a more complex history of science perspective, see Moro Abadía/Pelayo 2010. In 2016, the film *Altamira* starring Antonio Banderas and directed by Hugh Hudson, premiered in Santander, near Altamira. The film narrates the traditional view of the Altamira story.

24 Several books, documentaries and academic works have dealt with the Taung story. From the perspective of this book, the most useful point of reference is Richmond 2009.

25 Richmond 2009, 30–31.

These controversies and their main actors, Altamira and Sanz de Sautuola as well as the Taung Boy and Dart, proved to be crucial points of reference for the Orce Man actors, and especially for Josep Gibert, who used both controversies and their main characters as a model of how scientific controversies work and how the scientific community reacts to apparently controversial or innovative claims. Gibert often summoned Dart or Altamira to state that "what happened to them is what is happening to me".

Despite the study of scientific controversies being one of the main fields of research for social historians of science, so far, there is not much historiography entirely devoted to analysing this phenomenon.[26] Yet, for the sociology of scientific knowledge, controversies are a crucial field of study as they clearly and more openly reveal social aspects of scientific processes that usually remain hidden and inexplicit.[27] At the same time, studying controversial scientists, the anti-heroes in scientific controversies, also helps to reveal the ways that science works, as those marginalised by their peers have to 'sell' themselves and their positions more conspicuously. This situation brings the motivations and strategies of these scientists more clearly out into the open, again helping us understand the broader aspects of how scientific processes unfold.[28]

There is, in fact, a great deal of historiography of controversies linked to sciences and scientific issues that are similar or related to palaeoanthropology; for example, archaeology, geology or palaeontology. In this regard, the key work is Martin Rudwick's classic study *The Great Devonian Controversy*, which has been a great inspiration for this book.[29] Rudwick's chronological narrative or how he managed, in his own words, to make "small facts speak to large issues" is what this book also aims to achieve. Despite the lack of comprehensive analyses of how controversies develop in these sciences, many other case studies have been helpful for this research.[30]

Beyond controversies from geo-, palaeo-, and archaeo- sciences, there are several historical and sociological studies on scientific controversies. These

26 Volumes devoted to controversies include Engelhardt/Caplan 1987; Machammer/Pera/Baltas 2000; and Martin 2014.

27 Pinch 1994, 88.

28 For instance, Latour 1987, yet Latour's research does not deal with the role of mass media in these processes.

29 Rudwick 1985.

30 For instance, Clemens 1986, on the way that scientific debate shaped public debate; De Bont 2003, for a wonderful example of the strategies and discourses used by scientists; Pruitt 2011, for a detailed analysis of how supposedly 'bad' science attracts the public's attention and Wagenknecht 2012, who discusses how public appearances influence daily scientific practices.

studies have been very useful models in understanding and characterizing issues such as the necessity of scientists to 'go public', for funding and prestige, the use of the general media as a source of information for scientists, the way in which disputes helped disciplines to establish themselves and young researchers to foster their careers, the concept of 'tools' to refer to the resources used by actors in such controversies or how the end of a dispute is not a particular moment but often a process.[31]

Because of its similarities with the Orce story in terms of time frame, public presence and dispute among scientists, the Cold Fusion controversy deserves separate commentary, as it will be used as the reference for many of the historiographical categories used for the Orce Man. In 1989, Martin Fleischmann, a reputed electrochemist, and Stanley Pons, also an electrochemist, reported that they had registered excess heat in their electrolysis of heavy water in a palladium surface experiment. The results were first announced at a press conference, as they raised hopes for a cheap source of energy. Many scientists around the world tried to replicate the experiment, mostly with negative results and with the reporting of errors in the original experiment. Some months later, most scientists did not accept Fleischmann and Pons' claims and they received harsh criticism for both the experiment and the way they presented it to the world. Several sociologists and historians of science have used the so-called Cold Fusion saga to analyse the relations between science and its publics in the late 1980s and early 1990s. Although the case of the Orce Man deals with a different scientific discipline, it has several parallels with the Cold Fusion saga. Many of the analytical tools developed in the study of the Cold Fusion saga therefore will be used and applied in this book for the Orce Man case[32]

Despite being a growing discipline that has accumulated a significant amount of scholarly work, the history of palaeoanthropology still lacks a comprehensive synthesis that would allow it to play a role in the broader history of science studies.[33] In fact, palaeoanthropologists themselves have,

31 Works include controversies such as the Cold Fusion controversy, see next footnote; the Big Bang vs. Steady-state dispute, see McConnell 2000; Gregory 2003; Gregory 2005; Bucchi 1998, 82–100; or the Fluoridation controversy, see Martin 1991 and Martin 2014. The concept of 'medialisation' is also useful to understand this relation between science, controversies and the public, Rödder/Franzen/Weingart 2012 and Nieto-Galan 2011, 244.

32 Some of the academic works devoted to the Cold Fusion saga are Lewenstein 1992; Lewenstein 1995; Bucchi 1998, 36–81; and Gieryn 1999, 183–232.

33 Goodrum 2000; Goodrum 2009; and Gundling, 2010; and Delisle 2012 are the only attempts to develop useful and complete accounts of what the history of palaeoanthropology looks like. Yet, none of them analyses or deals with the relationship between palaeoanthropology and popular media.

almost since the very beginning of the discipline itself (from Raymond Dart's *Adventures with the missing link* to Donald Johanson's *Lucy. The beginnings of humankind,* down to the Spanish Atapuerca research team, or *A new human*, the story of the Flores 'hobbit' finding), recounted their own history and often the history of those before them, in order to achieve very concrete goals in their professional careers.[34] Usually in an easily accessible style, scientists and also science writers have published very successful books mainly focused on controversy or on the adventurous side of palaeoanthropology.[35] The Orce Man story fits into the wider history of palaeoanthropology historiography by providing a 20th century controversy case in Spain that helps to understand better the changes in palaeoanthropological science, its ways of creating knowledge, its relations with the public and politicians, from the African 'bone fever' of the 1960s and 1970s to the 'revolutionary' but also controversial discoveries in the early 2000s, such as the Flores 'hobbit', *Homo naeldi* and the Denisovans.[36]

Historians of archaeology or anthropology, sociologists and historians of science have also become interested in the history of human origins research. This kind of work has explored the influence of national identity, colonialism and politics in the interpretation of archaeological and palaeontological remains. Often this research shows how the social, cultural, political, historical and scientific context can explain these interpretations of bones.[37] In a similar vein, several scholars have devoted their work to specific cases such as the Peking Man, very convincingly showing its link to the development of Chinese national identity.[38] Recently, much has been written on Raymond Dart, the *Australopithecus* discovery and the subsequent controversy. Richmond was informed by recent studies on experts and credibility in science, Kuljian related this discovery to the concepts of race in South Africa and Madison highlighted the role of materiality and place in the controversy.[39] Murray Goulden has analysed the Piltdown case and the Flores 'hobbit' discovery by exploring

34 Dart 1959; Johanson/Maitland 1982; and Morwood/van Oosterzee 2007. For an analysis of these aims, see: Hochadel 2013b, 241–245.

35 Prime examples include Lewin 1997; Aczel 2007; Gibbons 2007; and Pyne 2016. Others on the Piltdown case are Spencer 1990; Walsh 1996, or the Peking Man case with Shapiro 1974.

36 Lewin 1997; Goulden 2013; Hawks et al. 2017; and Warren 2019.

37 Seminal works are those of Hammond 1982; Trigger 1992; Schlanger 2002; and Sommer 2007. Studying Spanish stories, we can find among others Díaz-Andreu 2012; Ruiz-Zapatero 1996; Ruiz-Zapatero/Castaño 1999; Moro-Abadía 2007, Moro-Abadía/Pelayo 2010; Lanzarote 2013; and Hochadel 2013d.

38 Sautman 2001; Leibold 2006; Schmalzer 2008 and Manias 2014.

39 Richmond 2009; Kuljian 2016; and Madison 2019.

the separation between science and popular culture using Thomas F. Gieryn's concept of 'boundary-work'.[40] For Gieryn, boundary-work is the process by which boundaries and demarcations between different kinds of knowledge, for instance, scientific and non-scientific, are created, attacked or reinforced.[41] This concept has become very useful in characterising the strategies used by different actors in the Orce controversy to exclude the work and research of a specific scientist from science itself, in what Gieryn calls the 'expulsion' type of boundary-work, a process closely related to the loss and acquisition of credibility.[42]

The ways and spaces in which the credibility of the Orce fragment was built and supported, its changing interpretations over time and the processes by which the credibility of a claim or a scientist were built are among the main interests of this book. Also within its scope is to consider how actions taken by scientists and claims presented were initially accepted by society and the other scientists before the controversy, only to be rejected once it started. Why a claim was accepted at first and was later deemed not to be credible is analysed in detail.

The controversial bone travelled from the excavation to the research centre and from there to another research centre in a foreign country. Then it stayed for years in a bank vault.[43] At some point in the story, it was publicly exhibited and even touched by journalists and other researchers, but it was also kept away from the public eye and access was sometimes restricted to certain researchers. The actors who controlled the bone, or had the legitimacy to write about it, controlled the excavations and therefore, further research in the area.

During the research for this book, I had the opportunity to analyse the archives of the main character of this story who died just two years before I began my work. In his office, among hundreds of letters, research material and draft papers, Gibert had also kept several newspaper clippings about the story, including photocopied newspaper pages and some even sent by fax. When I interviewed another main character of this story, he lent me a folder containing quite a few relevant documents. Again, among letters, excavation permits and other institutional papers, there were several newspaper clippings. Finally, sometime later, I interviewed a young geologist who was also a participant in the story and who gave me, along with several published scientific papers and video recordings, a CD with a large number of online versions of newspaper articles on the most recent part of the story.

40 Goulden 2009a; Goulden 2009b; and Goulden 2013.

41 Gieryn 1983 and Gieryn 1999.

42 Gieryn 1999, 15–16.

43 For more on the situation of the Orce bone during the first controversial years, see Annex 3.

This shows how the scientists involved were not only aware of the public dimension of the story but also gave it a great deal of significance in the story's development, as a source of information and even in their claims. What was said in the newspapers was, for these scientists, almost as relevant as what was said in the scientific literature. In the case at hand, the 'public perception' of a specific scientific claim was as important as the 'scientific perception' of it. At times, these two perceptions appeared completely intermingled and very difficult to separate. In short, this story is ideal for exploring the interactions between scientists and their 'publics', including journalists, other scientists, the general public and politicians.

The reasons for this story's being so public are manifold. Perhaps the main reason lies in the general characteristics of palaeoanthropology itself. Palaeoanthropology and human origins research in general are sciences that cannot promise any material or practical benefit to politicians, policy-makers or industry.[44] However, the quest for our ancestors' remains attempts to answer profound questions about who we are and where we come from. These rather philosophical or even theological questions, their possible answers and especially the discovery of yet another 'missing link' never fail to fascinate and capture the imagination of the general public.[45]

Historians and sociologists of science have always tried to understand the processes by which knowledge is created and disseminated beyond the traditional view of a simplification of content from the closed space where scientific knowledge is created to the open arena of the public domain. In the 1970s, with increasing interest in the social processes of science, some scholars focused their attention on the role of what is called 'popular science' or the 'popularisation of science'. The general public emerged as a key actor in the construction of scientific facts.[46] One of the models proposed to explain this relationship established that it is often impossible to distinguish clear boundaries between 'pure science' and its 'popularisation'. Therefore, it would be better to talk about a 'continuum'. Others even talk about a web, where knowledge flows between different genres of science communication made up of different categories of texts which cannot be fully separated from each other.[47]

These models have shown how the boundaries between scientific research and its public dissemination are more fluid than previously thought and how

44 Bucchi 1998, 93, indicates a similar case for cosmology.

45 Kjærgaard 2011b.

46 For instance, Shinn/Whitley (eds.) 1985. The rediscovery of German scholar Ludwick Fleck was crucial to this turn to popularisation. On Fleck, see Nieto-Galan 2011, 303–309.

47 Hilgartner 1990 and Lewenstein 1995.

popularisation must be considered not only as a simplification of scientific knowledge but as a part of its construction. Yet, going beyond the categories of 'science' and 'popularisation', James A. Secord proposed considering that knowledge is made 'in transit'. For him, to study science the historian should study everything that is related to its communication, from scientists to publishers, to bookbinding and prices.[48]

Relating palaeoanthropology to mass media, Marianne Sommer analysed the role of newspaper representations of Boule's Neanderthal, concluding that Boule used "the press as a means to get public sympathy and increase acceptability for their scientific views".[49] In 2011, Peter C. Kjærgaard published an article on the worldwide presentation of two recent findings: Ida, 'the missing link' fossil and the famous *Ardipithecus*, both presented to the world in 2009. In his analysis, Kjærgaard states that the way that these findings were announced makes it difficult to distinguish between "hype, advertising, science communication, and well-meaning public engagement".[50] In 2016, I participated in a volume of *Centaurus* that used different case studies from the late 19th century and all of the 20th century to analyse the relation between the creation of prehistoric knowledge and newspapers all over the world.[51] Most of these cases clearly showed the interest that the researchers themselves had in what was said in the press, a very similar instance of what happened in the Orce Man case.

Among the three historiographies reviewed – the study of scientific controversy, the history of palaeoanthropology and the relation of science and its publics – there is a clear overlap. Situated within this overlap, the Orce case will show that the mass media became the stage on which the scientific debate took place, increasing science's 'publics'. The impact of this process on the political side of science, on the negotiation of the 'boundaries' of science, on the way in which credibility was built and the processes of closure in the scientific controversies themselves will also be at the forefront of the analysis.

Following the three historiographies, this book aims to study the Orce Man controversy considering not only what appeared in scientific papers, conferences and other forums, but also and especially, what was said in the public arena. This public discourse, this book will argue, was not only different from the scientific one, but was in fact, an important part of it. During the Orce Man controversy, the scientific, the political and the public became in many

48 Secord 2004, 655 and Secord 2002.
49 Sommer 2006, 233.
50 Kjærgaard 2011a.
51 Hochadel/Carandell/Florensa 2016.

instances many facets of the same thing. Finally, this book will show how the Orce Man, the bone itself, was not an objective, independent thing taken from nature, but something different: it had several interpretations, appropriations and even uses for the different actors involved. From its public presentation, the bone (figure 1) became much more than a hominid or equine cranial fragment; it was the flagship for certain claims, a way to control excavation sites, a gain for the next election, a way to get credibility restored, the solution to economic problems and even a chance for public recognition. Getting to grips with the Orce Man story will therefore mean tackling all of these issues and will lead us to better understand many other potential Orce Men.

CHAPTER 1

Discovery (1976–1984)

On 11 July 1983, three young scientists had an astonishing experience. They were the centre of attention of a highfalutin press conference in Granada with several journalists, seasoned scientists and high-level politicians. The spotlight shone on them as the brilliant discoverers of a tiny cranial fragment of the first inhabitant of Europe, apparently turning palaeoanthropological science on its head. At the same press conference, an agreement was signed guaranteeing funding for the young researchers and their institution. Afterwards, their finding, the Orce Man, became known to everyone in early democratic Spain.

The Orce Man was discovered in a very concrete scientific context of the hunt of the First European of which these young researchers were part of. This will lead to the introduction of the main characters, the scenarios of the discovery, and the political context. Later, a chronological account of the early phases of the Orce story will be provided, from the discovery itself to the first signs of a controversy breaking out, with special attention being paid to media reactions. At this point, the Orce Man was a successful and impressive discovery that amazed and interested Spanish society. The context of post-Francoist Spain and the ongoing state building process will emerge as crucial to the way the discovery was treated and brought into the public arena by scientists, politicians, and journalists.

Likewise, the geographical location of the action, the bone and the actors will also be revealed as a prominent factor for understanding the way that the interactions between them worked. The material control of the Orce Man will also emerge as a main issue during its public presentation. Finally, the 'science and the media' phenomenon and what the discovery of the Orce Man can tell us about it will be analysed.

1.1 Setting the Scientific and Political Scene

In the late 1970s and early 1980s it was quite clear that humanity had originated somewhere in Africa around 3.5 million years ago. The time and route used by the descendants of these hominids in their migration to Europe was, however, not so clear.[1] Although different scientists' interpretations in palaeoanthropology hardly coincide, according to the majority of scientists opinions, at the

1 Phillips 1980, 37.

 | DOI:10.1163/9789004431508_003

time, the oldest hominid fossil remains in Europe were two skull fragments that can be seen as an old friend and an amazing newcomer. On the one hand, the Mauer mandible found in 1907 near Heidelberg in Germany, which was dated at about 400,000 years old and was still considered by many as the representative of the first species (*Homo heidelbergensis*) to come to Europe.[2] On the other hand, the Tautavel Man found as recently as 1971 in the Arago Cave in Tautavel, a small town in Southern France, near the Spanish border. It was a very impressive and almost complete 'pre-Neanderthal' cranium, as the discoverers called it.[3] Marie-Antoinette and Henry de Lumley were the discoverers of the Arago hominid that was dated then at approximately 330,000 years old.[4] Ever since, and due to the de Lumleys' great popularisation efforts, Tautavel was transformed into the "home of the First European". At that time, Henry de Lumley was a very powerful figure in French prehistory and palaeoanthropology. He was, since 1980, professor at the *Muséum National d'Histoire Naturelle*, director of the prehistory laboratory in the *Musée de l'Homme*, and director of the *Institut de Paléontologie Humaine* (all three institutions located in Paris). Henry and Marie-Antoinette were famous for the discovery of the Tautavel Man, but also for the excavations in the Vallonnet site, the Terra Amata site, and the Lazaret cave. The de Lumleys also carried out intense outreach work related to these sites, work that led, for instance, to the opening of a museum in Tautavel, a Terra Amata museum, the direction of various temporary exhibitions, and later also popular science books.[5]

These were the two main oldest hominid remains, but a different issue was the earliest evidence of hominid presence. By the mid-1970s, after the Tautavel find, Henry de Lumley and the Catalan archaeologist Eudald Carbonell (a disciple of de Lumley) started a 'hunt for the First European' that placed hominid arrival to the continent at around 1.5 million years ago, following the statements of Czech prehistorian Karel Valoch (1920–2013).[6] Valoch claimed in 1976

2 Phillips 1980, 38.

3 AMC-ICPMC: Letter from Crusafont to de Lumley, 3 August 1983.

4 De Lumley/de Lumley 1971, and de Lumley et al. 1977. See also Chippindale 1986. The Petralona skull in Greece and the remains in Vértesszöllös in Hungary were also considered some of the earliest hominids in Europe, but in any case neither was more than 500,000 years old. Phillips 1980, 38.

5 For de Lumley see: Wikipedia Contributors 'Henry de Lumley'. For Terra Amata and Tautavel: Wikipedia Contributors 'Musée de paléontologie humaine de Terra-Amata' and Anonymous 'Un peu d'histoire …'. See also the exhibitions: de Lumley 1982a and de Lumley 1982b, which toured some cities, including Barcelona, Diputació de Barcelona 1985 and Amela, *La Vanguardia* 1985, and the popular science books de Lumley 1998, de Lumley 2007, and de Lumley/de Lumley 2011.

6 Hochadel 2013a, 400.

that stone tools revealed that hominids had lived in the Sandalja Cave in Istria 1.5 million years ago and that consequently, at that time, the site had to be considered "the oldest Palaeolithic site in Europe".[7] In 1981, de Lumley organised an exhibition in the *Musée de l'Homme* in Paris with a title that perfectly reveals the chronology of hominid arrival to Europe that he was positing: *Les premiers habitants de l'Europe: 1.500.000 – 100.000 ans*.[8]

In addition, the book *Prehistoric Europe*, published in 1984 as an "elementary textbook to teach European Prehistory", stated that the "chronology for the colonisation of the continent put forward by Henry de Lumley" situated stone tools from the Vallonnet site (southern France) at "between 0.9 and 0.95 million years old".[9] According to de Lumley, this site "changed everything" regarding the notion of the 'First European', becoming an important site for the his 'hunt'.[10] But de Lumley and Carbonell were not alone. Another French scientist, Eugène Bonifay, was independently claiming the discovery of even earlier stone tools (up to 2 million years old) in some sites in the French Massif Central. In an article in the popular French science magazine *La Recherche*, Bonifay claimed that 'Les traces des premiers Hominidés en France' were at least 1 million years old and possibly up to 1.8 million years old.[11] Josep Gibert, one of our main actors, had both de Lumley's and Bonifay's publications, which leads us to think that even if he had not yet found any hominid remains, he was certainly following debates around the First European settlement.

Closer to the time of our story, in November 1982, an Italian research group published an article in *Nature* on the "abundant stone tool industry" dated at 0.7 million years old at the Isernia – La Pineta site, between Rome and Naples. According to the researchers, with this dating the site became "one of the most important localities for the question of the earliest human colonisation of Europe".[12] Later, the Orce Man discoverers mentioned this site in the press and in their early scientific articles.[13]

To sum up, although among the international scientific community no hominid remains of more than 400,000 years old were accepted in Europe, several scientists claimed there was an earlier occupation, using stone tools as their main evidence. The settlement of Europe was an on-going issue, a source

7 "le gisement paléolithique le plus ancient en Europe", Valoch 1976, 178 and Phillips 1980.

8 Hochadel 2013b, 80; de Lumley 1982a; de Lumley 1983.

9 Champion et al 1984, v and 27. De Lumley 1976.

10 De Lumley 2007, 13 and 112–113.

11 Phillips 1980, 37–38; Champion et al. 1984, 27; and Bonifay 1981, 1442–1444.

12 Coltorti et al. 1982.

13 For instance: De Semir, *La Vanguardia* 1983 and Agustí/Gibert/Moyà-Solà 1983a, 8–9.

of debate. In the early 1980s, when our story takes place, the scientists that supported this claim appeared to be looking for definitive confirmation. It is in this scientific context of the quest of the 'First European', that the discovery of the Orce Man must be situated and understood. Additionally, it is worth highlighting how emphasis on the 'possession' of key sites is frequent in palaeoanthropological and archaeological scientific and popular publications. Researchers often emphasise how they have found the oldest and most crucial hominid remains or stone tools, usually stressing over and over again how these new discoveries mean great shifts in scientific ideas and conceptions. This, of course, also entails a significant popularisation effort in the scientists' pursuit of recognition and funding.[14]

In the summer of 1976, a little expedition from the *Institut de Paleontologia*, based in Sabadell, Catalonia, arrived to the Guadix-Baza geological depression near Granada, Andalusia. Following the remains of an ancient lake formation, this expedition found what seemed to be an important site in a location called Venta Micena in the municipality of the rural town of Orce, which had less than 2,000 inhabitants at the end of the 1970s.[15] The *Institut de Paleontologia* had emerged from the Palaeontology Department of the *Museu de Sabadell*, situated in the more than 150.000 inhabitants industrial city of Sabadell, thirty kilometres from Barcelona (see map 1). Its main promoter was Miquel Crusafont i Pairó who was a prominent character in Spanish palaeontology during Franco's dictatorship.[16] A skilled Miocene palaeontologist, with connections to leading international scientists, Crusafont was a dominant figure who decided the PhD topics of his disciples with the goal of reviewing the taxonomical groups of all Iberian mammal fossils.[17] Among Crusafont's last pupils were three of the main characters of our story (figure 2): Josep Gibert i Clols, who received his PhD on insectivore fossils in 1973; Jordi Agustí i Ballester, who in 1981 completed his thesis on rodent fossils; and Salvador Moyà-Solà, who in 1983 finished his own thesis on bovine fossils. All three were collaborators of the *Institut* (Gibert and Agustí even organised and participated in the expedition that found Venta Micena), but none of them had a permanent position.

14 For several examples of palaeoanthropological research with these features, Fahnestock 1989; Lewin 1997; Gibbons 2002 and 2007; and Hochadel 2013a and 2013b.

15 Gibert 2004, 23–25. For Orce inhabitants: Wikipedia Contributors 'Orce'.

16 Florensa 2013; Florensa 2017 and Acosta 2013.

17 Gibert 2004, 22; Interview with Agustí 2012; and Català-Gorgues 2013.

FIGURE 2 From left to right: Salvador Moyà-Solà, Jordi Agustí, Josep Gibert and a young Lluís Gibert at the *Institut de Paleontologia* in 1983
FROM J.A., *DIARI DE SABADELL* 1983B

Moyà-Solà had a PhD fellowship, Agustí also had a fellowship while working as a secondary school teacher, and Gibert, the most senior of the three, was also working as a secondary school teacher while giving some lectures at the *Universitat de Barcelona*, where Crusafont was professor.[18] At the beginning of the 1980s, Crusafont remained director of the *Institut* despite being an elderly and seriously ill man. He barely went to the *Institut*, which, beyond collaborators, only had one other full-time permanent scientific employee.[19] After the 1976 expedition, different small excavations were conducted in Venta Micena. Gibert, Agustí, Moyà-Solà, the amateur palaeontologist Joan Pons Moyà, and Gibert's son Lluís (who was very young at that time) participated in these excavations, which seemed to support the palaeontological importance of the site (figure 2).[20]

18 Interview with Jordi Agustí 2102; Interview with Salvador Moyà-Sola 2012 and Gibert, 2004, 33.

19 This single scientific employee was Josep Vicenç Santafé i Llopis (b.1934–2017), also a disciple of Crusafont's who later specialised in dinosaur tracks. JagBa 'Josep Vicenç Santafé i Llopis'.

20 Gibert 1999a; Interview with Agustí 2012; and Interview with Moyà-Solà 2012.

The *Instituto Provincial de Paleontología de Sabadell*, later *Institut de Paleontologia de Sabadell*, was created in 1969 under the sponsorship of the *Diputación Provincial de Barcelona* (Barcelona's provincial council) which had economic and political control over it.[21] In 1969, Spain was under Francoism, a fascist government very centralised in Madrid and led by the dictator Francisco Franco.[22] The *Diputaciones Provinciales* were the only 'regional' institutions allowed by the central government and were used to control the different parts of Spain. After Franco's death in 1975, Spain entered a period called the 'Transition' in which the country moved into democracy leaving behind the past in all its senses. At the same time, Spain also went from being a centralised state to a state divided into 17 *autonomías* (autonomous communities) that had some degree of political control. Catalonia and Andalusia were two of these *autonomías*.[23]

In 1980, the *Generalitat de Catalunya* (the recovered Catalan autonomous government that joined the Barcelona, Girona, Tarragona and Lleida *Diputaciones Provinciales*) started its first parliamentary term under the right-wing Catalan nationalist party *Convergència i Unió* (CiU). The second most voted party was the Catalan branch of the Spanish socialist workers' party (*Partido Socialista Obrero Español* – PSOE), the *Partit Socialista de Catalunya* (PSC-PSOE). Finally, the third parliamentary force was the communist party, *Partit Socialista Unificat de Catalunya* (PSUC). At the same time, the PSC-PSOE won local elections (in both 1979 and 1983) in Barcelona and most of the cities in the Barcelona province, while PSUC won in Sabadell. This factor gave the PSC-PSOE control of the *Diputació de Barcelona* (now with the new Catalan name: from *Diputación Provincial* to *Diputació de Barcelona*). After its dominant position during the dictatorship, the *Diputació* was being reorganised with no clear power in a new relationship with the Catalan autonomous government and in the context of an emerging democracy.[24] As Antoni Dalmau, lawyer and head of the *Diputació*, acknowledged, in 1983, the role of the *Diputació* was undefined.[25]

In October 1982, the socialist Felipe González (b. 1942), from the PSOE, won the Spanish general elections for the first time for a left-wing party, after forty years of dictatorship. In Andalusia, the PSOE also won the first elections for the

---

21 For the history of the *Institut*, Acosta 2013, 337–340.

22 For a good overview of Francoism and Franco see: Molinero/Ysàs 2008.

23 For the history of the *Diputació de Barcelona* during Francoism see: Molinero/Ysàs 1988. For an account of the Transition to Democracy see Molinero/Ysàs 2018.

24 Rius/Martínez 2012, 13–15 and Mascarell 1988, 125, 129.

25 Dalmau, *La Vanguardia* 1984.

Junta de Andalucía, the recently created autonomous government. The new *autonomías* were in the process of 'construction' and needed legitimisation actions that would help to strengthen them. Considering this political context, it is important to highlight a couple of crucial points here. The first point worth emphasising is that Josep Gibert, in the late 1970s and early 1980s, was an active member of the communist party PSUC. He participated in the first municipal elections in a little town near Sabadell, Castellar del Vallès, where he lived. Later, in 1986, Gibert also participated in the general elections as number 10 on the list of the newly formed *Partit dels Comunistes de Catalunya* (PCC), a pro-Soviet party that had separated from the PSUC.[26] The second point is that while the Catalan autonomous government was controlled by the right-wing CiU, the Catalan socialists controlled the *Diputació de Barcelona,* which was used as an instrument by the PSC-PSOE to maintain political power at a level between the town councils and the autonomous government. In addition, tensions between socialists and the right wing were especially complex at that time.[27] The situation in the Spanish central government gave power to the PSC-PSOE, but it still needed to reinforce the *Diputació* to highlight its role and the usefulness of its actions.

What to do with the *Institut de Paleontologia* was one of the many decisions that the *Diputació* had to take. As mentioned earlier, the *Institut*'s founder and director, Crusafont, was old and ill and therefore the *Diputació* had to decide who would be his replacement and what to do with the scientists that worked within the *Institut* and their precarious situation. Therefore, along with the future of the *Diputació*, the future of the *Institut de Paleontologia* was also being decided.

In 1981, Moyà-Solà and Agustí co-authored an article with Eudald Carbonell and Pons Moyà, among others, in *Endins,* the journal of the *Federació Balear d'Espeleologia* (Balearic Speleology Federation). In it, they stated that Cueva Victoria, a site in Murcia discovered by Pons Moyà some years earlier, was the oldest site with human occupation on the Iberian Peninsula (map 1). The authors dated Cueva Victoria at between 0.7 and 1.5 million years old and compared it with several European sites, including de Lumley's and Bonifay's.

---

26 For the municipal elections in Castellar, see AJG-ICPMC: Letter from Gibert to Narcís Serra, 15 April 1996. For general elections with the PCC see: Redacción, *El Periòdico* 1986a. Among Josep Gibert's papers there were a couple of issues of *Realitat,* the PCC's magazine. Finally, for some comments on politics from Gibert himself see: Interview with Josep Gibert 2003.

27 Rius/Martínez 2012, 15 and 19–21. Mascarell 1988, 129.

They also considered Venta Micena, the Orce site discovered some years before, to be a little older than Cueva Victoria but claimed that the former did not present clear human presence.[28] Although this Cueva Victoria finding was not reported on much in the Spanish media, it shows how at least two of the Orce discoverers, Agustí and Moyà-Solà, were active participants in the 'hunt' for the 'First European'. In fact, before the summer of 1982, the third of the trio, Josep Gibert, stated to the Sabadell local newspaper, *Diari de Sabadell*, that Venta Micena seemed to be a potential site for finding a hominid fossil.[29] In a later account of the period of the late 1970s and early 1980s, Carbonell and Moyà-Solà explained how one day, coming back from Cueva Victoria, dirty and tried, they began to drink and promised each other that they would find "the oldest remains ever found".[30]

In the summer of 1982, Gibert organised an educational excavation summer camp led by Moyà-Solà, Agustí, and himself, gathering several secondary school students to excavate at the Venta Micena site. One of the students was Bienvenido Martínez-Navarro, who would later become another main actor in this story.[31] During the excavation, two students unearthed a piece of bone that for the site leaders seemed to come from a hominid.[32] The bone was stuck to a rock and thus Moyà-Solà decided to extract the whole block and take it back to Sabadell for a more accurate cleaning process.[33] The Orce Man was being discovered.

1.2 'Look What We've Found!' The Orce Man among Politicians, Experts and the Public

Around Christmas 1982, Salvador Moyà-Solà cleaned the extracted block in the *Institut de Paleontologia de Sabadell*. The piece turned out to be a cranial fragment of less than 10 cm with parts of the parietal and occipital bones.[34] After this first preparation by Moyà-Solà, most of the inner part of the cranium was still attached to the rock and was impossible to remove without breaking the

28 Carbonell et al. 1981.

29 Ache, *Diari de Sabadell* 1983.

30 "els ossos més vells mai trobats", Antón, *El País* 2000.

31 Martínez-Navarro 1993, 13.

32 Gibert 2004, 30 and J.A., *Diari de Sabadell* 1983a.

33 Interview with Agustí 2012 and Interview with Moyà-Solà 2012. For more on the travelling character of the bone, see Annex 3.

34 For more details, see Annex 1.

bone. More appropriate equipment was necessary, and, at that time, the *Institut* did not have it.[35] Therefore, and crucially, the inner part of the bone was not visible (see figure 3). The researchers had doubts about the actual classification of the bone but it seemed that the wide cranial curvature and a comparative anatomical differential diagnosis conducted by Moyà-Solà (the only one of the three that had expertise in macrofauna) allowed them to classify it as hominid.[36] Beyond the obvious concerns that such a tiny and incompletely visible fragment could raise, according to later accounts of the researchers, it seemed that many doubts emerged around its dating.[37] The deposit in which it was found was dated by Agustí using biostratigraphic methods at approximately 1.4 million years old, or at least between 0.9 and 1.6 million years.[38] These dating results implied that the Orce bone could be the oldest European hominid remain. In spite of this, for many (including the discoverers), the Orce finding did not mean a great conceptual shift, but the confirmation of the hypothesis of an early hominid presence in Europe. The bone could cause a 'revolution' because it could be the oldest hominid remain in Europe but not because it shifted conceptions about hominid arrival to Europe, as was often stated later by different parties.[39] A considerable group of scientists were claiming hominid arrival to Europe to be around 1.5 million years ago, despite, of course, this being a controversial claim for many others.[40] In any case, in February 1983, Gibert received a personal letter in which the correspondent, a friend of the family, congratulated him on the "magnificent discovery".[41] Therefore, by the beginning of 1983 they were sure that the Orce bone was an important finding.

Sometime before May 1983, the discoverers decided to bring the bone to Eduard Porta i Ferrés, museologist, head of the Museums Department of the *Diputació*, on which the *Institut* depended. Moyà-Solà, Agustí and Gibert probably went to the nowadays Palau de la Generalitat, the main Catalan government building in the very central Barcelonese Plaça Sant Jaume, that at that time was the headquarters of the *Diputació*. Porta received the finding with an eager "Fantastic!"[42] As mentioned earlier, the *Diputació* was undergoing a process of redefinition that would allow it to validate itself and survive within the

35 Interview with Moyà-Solà 2012.

36 Interview with Agustí 2012; Interview with Moyà-Solà 2012; and Gibert 2004, 33.

37 Interview with Agustí 2012.

38 Gibert/Agustí/Moyà-Solà, 1983a and Interview with Agustí 2012.

39 Gibert 2004, 52 and Interview with Agustí 2012.

40 See a review of this dispute in Dennell/Roebroeks 1996.

41 "Descubrimiento magnífico", AJG-ICPMC: Personal letter from an unknown sender to Gibert, 9 February 1983.

42 Gibert 2004, 32 and Interview with Agustí 2012.

FIGURE 3 The cranial fragment with the inner part still attached to the rock. Left, external part; right, inner part
FROM CAMPILLO 2003, SHEETS 9 AND 10. PHOTO ORIOL CLAVELL

new democratic state. The discovery of the Orce fragment could thus be one step more towards the establishment of the *Diputació* as a worthy institution paying special attention to history, culture, and heritage.[43] On the discoverers' side, this 'look what we've found' strategy seemed to show how scientists were seeking better funding conditions for the *Institut* and for themselves within the on-going renovation of the *Diputació*. As Gibert himself acknowledged, "my status in the *Institut* was precarious, as was Agustí's and Moyà's, since there wasn't any formal link, only a scientific collaboration, which in my case had already lasted sixteen years, since 1967".[44] When the discoverers went to see Porta, he recommended they bring the bone to Domènec Campillo Valero, a well-known neurosurgeon and palaeopathologist, to further confirm the hominid nature of the remains.[45]

43 Rius/Martínez 2012.
44 "Mi posición en el Instituto de Paleontología, y también la de Agustí y Moyà, era precaria, pues no había ningún vínculo administrativo sólo de colaboración científica. Colaboración que en mi caso duraba ya dieciséis años, desde 1967" Gibert 2004, 32.
45 Gibert 2004, 32; Campillo, 2002, 23, 25; Interview with Campillo 2011.

For Campillo, the Orce bone was indeed human due to its curvature and thinness: there was no doubt, no possible mistake (figure 3).[46] With Campillo's support, the scientists returned to Porta who encouraged them to seek international confirmation.[47] To do so, the discoverers went to Lyon to see Pierre Mein, a French palaeontologist who had a very good relationship with Crusafont and his students. In Lyon, Rafael Adrover, a Spanish palaeontologist who also had ties with the *Institut*'s scientists, saw the bone too. Both stated that it belonged to a hominid.[48]

Back in Sabadell, Peter Andrews, a British palaeontologist from the *Natural History Museum* who was there doing research on the *Institut*'s dryopithecines, also saw the specimen. Andrews too confirmed that the bone was hominid and recommended an international publication.[49] During the popularisation of the finding, all of these experts were presented as the international confirmation of the discovery's validity.[50] Later, they were also very often presented by Gibert as the international endorsement that allowed the discovery to be announced.[51] Yet, it seems that at least Andrews visited Sabadell just a week before the public presentation of the bone fragment in June. Andrews thus did not provide confirmation before publishing the finding and deciding to popularise it; he was not an international endorsement, just very welcome news when everything was already decided, the publications were already prepared, and public announcements were about to be made.[52]

The way that Moyà-Solà brought the fragment from Orce to Sabadell to clean it up; the discoverers' doubts and the need for expert confirmation of the bone's actual validity as a human remain; how the three discoverers presented the fragment to the *Diputació* politicians ... These elements again stress how scientific discoveries must not be seen as single 'eureka' moments taking place at the site itself, but as complex processes of interaction between the scientific object, scientists, their instruments, and other agents such as politicians or external experts, where periods of uncertainty during investigations may easily arise. In addition, these early movements of the Orce bone and its growing relevance, show the importance of considering the materiality of natural objects

46 Interview with Campillo 2011 and Campillo, 2002, 25.

47 Interview with Agustí 2012. In his book, Gibert stated that seeking international confirmation was their idea, the discoverers' idea. Gibert 2004, 32.

48 For more on Adover: Moyà-Solà/Pons 2007. And on Mein: Wikipedia Contributors 'Pierre Mein'.

49 Gibert 2004, 32. More on Peter Andrews: Andrews 2011.

50 For instance: Agustí, *La Vanguardia* 1983.

51 Gibert 2004, 32.

52 Ache, *Diari de Sabadell* 1983.

and the role of the place where are found, studied and presented to different actors. This process shows how the Orce bone was transformed from a natural object into a very important scientific object with the intervention of the discoverers, but also other scientists, politicians and, as we will see, journalists and the public.[53]

Once the experts saw the bone, the three scientists went back to Porta. With the same enthusiasm, Porta led them to Antoni Dalmau, head of the *Diputació*. Dalmau then got in contact with the president of the *Junta de Andalucía*, Rafael Escuredo, since the bone had been found in Orce, in the Autonomous Community of Andalusia.[54] Escudero was governing Andalusia for the Andalusian PSOE, the same political party that controlled the *Diputació*. The presidents of the *Diputació de Barcelona* and the *Junta de Andalucía* agreed to organise a joint press conference to present the fragment to the public and to publicly sign a scientific agreement between both institutions regarding the Orce research.[55] The scientific collaboration between these two institutions would be presented as a model for collaborations in the newly democratic Spain.[56] Yet, the agreement obviously overlooked two other institutions, the Catalan *Generalitat* and the *Diputación de Granada* (Orce is located in the province of Granada). As stated earlier, the Catalan government was controlled by CiU, a right-wing political party.

In early 1983, the *Diputación de Granada* was governed by *Unión de Centro Democrático* (UCD), a political party that disappeared in February 1983. After the local elections in May 1983, the *Diputación de Granada* also came under the control of the PSOE.[57] Later, the newspaper *El Defensor de Granada* printed statements by the *Diputación de Granada*'s head of culture who highlighted that the *Diputación* had been "discriminated against" in the Orce issue and added that "this will not happen again" as the *Diputación de Granada* would be, from then on, "informed of all the research carried out and findings made in Granada."[58] It seems then that the fact that the same political party governed the *Diputació de Barcelona* and the *Junta de Andalucía* meant that the collaboration ignored both the Catalan *Generalitat* and the *Diputación de Granada*.

53 For more on accounts of discoveries see Woolgar 1976. See also Madison 2019. For more on where the bone was from its discovery to 1984, see Annex 3.

54 Interview with Agustí 2012 and Gibert 2004, 36.

55 Montero, *ABC* 1983.

56 De Semir, *La Vanguardia* 1983 and Redacción, *El País* 1984.

57 Lozano 'Elecciones Municipales en Andalucía 1979–2015'.

58 "Discriminada", "esto no volverá a ocurrir", and "informada de todas las investigaciones y hallazgos que se realizaran en Granada", Redacción, *El Defensor de Granada* 1983.

So what was presented as an exemplary scientific collaboration within the new democracy and its territorial organisation was in fact a collaboration based mainly on political affinities.

In their later accounts, both Agustí and Gibert stated that the snowball effect around the finding was intensifying and for them was at that point already unstoppable. According to both, they were young and inexperienced, and the active involvement of politicians led them towards a large public presentation that they never wished for.[59] According to these later remarks by Agustí and Gibert, the significance of the discovery was overemphasised, a statement that they never made back in 1983. With regard to the reasons for the subsequent media attention, Gibert and Agustí point to politicians that wanted to take political advantage of a scientific achievement and to journalists that, of course, jumped on board with vigorous enthusiasm.[60] However, the discoverers needed plenty of popular attention for the discovery to secure the *Institut*'s survival and their own professional future. So far, the Orce Man was a very successful scientific discovery, which attracted a great deal of public attention and created no problems or controversy. A great victory for the discoverers and for the new democratic state Spanish science.

Before the press conference, and maybe because it was scheduled, Gibert, Agustí, and Moyà-Solà published the bone as *Homo* sp. in May 1983 in a special issue of *Paleontologia i Evolució*, the journal of the *Institut de Paleontologia de Sabadell*. According to the discoverers, their early doubts no longer worried them because of the 'experts'' statements, despite the fact that at least Peter Andrews saw the bone only after this publication. This brief scientific article described the geological characteristics of Venta Micena and the anatomical features of the fragment, which they labelled VM-0, after Venta Micena, and a zero as it was the site's first hominid remain. Finally, they concluded that the fragment was the "oldest [hominid] in Europe and Asia". According to the three researchers, "this indicates that the human colonisation of Europe happened much earlier than assumed until now".[61] Not only politicians and journalists overemphasised the importance of the bone; the discoverers themselves had already attached a great deal of significance to their finding.

---

59 Bucchi 1998, 91, presents very similar statements from scientists related to the Big Bang-Steady State cosmology controversy. For a detailed study of the news devoted to the Orce Man see Annex 2.

60 Interview with Agustí 2012; Gibert 2004, 38; Martínez-Navarro 1993, 17.

61 "más antiguo de Europa y Asia", "indican que la colonización de Europa por el hombre fue mucho más temprana que lo hasta ahora supuesto", Gibert/Agustí/Moyà-Solà, 1983a, 8.

The *Junta de Andalucía* bought Venta Micena land from Tomás Serrano, owner of the property, before the press conference. Serrano appeared in several newspapers as the 'other' Orce man. Before the arrival of the *Institut* scientists, it seemed that he had already noticed the presence of fossils on his property and brought them to scientists in the *Museo Arqueológico de Granada*, but they ignored him. When the *Junta*, together with Gibert, bought the land, the significance of the discovery seemed to be hidden from Serrano, who later complained to the press: "If I had known [the importance of the discovery] the amount of money that they gave me would not have seemed fair to me. I've been poor all my life."[62] With the scientific publication and the site's ownership resolved, everything was set for the 'debut': the public presentation of the oldest European.

The first public news of the Orce discovery appeared in the local Sabadell newspaper *El Diari de Sabadell* a week before the press conference. On Saturday, 4 June 1983, it announced 'The remains of the oldest man in Europe have been found.'[63] The news piece featured statements by both Crusafont and Gibert and highlighted that the discovery had remained a secret so far because the *Diputació de Barcelona* and the *Junta de Andalucía* were negotiating a scientific agreement and had organised a press conference to present it. The next day, the Orce fragment began its journey through Spanish headlines. A press release, probably following the Sabadell newspaper, was submitted to the media by the EFE news agency, as an *ABC* piece indicates. This made the news pieces very similar across different newspapers.[64] The main Spanish national newspapers announced the forthcoming presentation with a very brief summary of the characteristics of the bone, its dating, and the crucial importance of the discovery in locating "the cradle of European and Asian men" and "the origin of the life of *Homo erectus*" in Andalusia.[65]

Yet, one section in *ABC*'s report was different. This newspaper highlighted that the *Junta de Andalucía* should make the necessary arrangements to "keep

62 "Si yo lo hubiera sabido, no hubiera sido justa la cantidad que me han dado. Yo he sido pobre toda la vida", W. L., *El País* 1983. For Serrano's story see: Gibert 2004, 37; Pedregosa/Ferreras, *El Periódico* 1983; Lyon, *El País* 1983b; and Chamorro 1983.

63 'Hallados los restos del Hombre más antiguo de Europa', Ache, *Diari de Sabadell* 1983, 3.

64 For instance, Montero, *ABC* 1983, and Redacción, *La Vanguardia* 1983a.

65 "la cuna del hombre europeo y asiático", Montero, *ABC* 1983. "El orígen de la vida de Homo Erectus en Andalucía", Redacción, *La Vanguardia* 1983a.

the discovery on Andalusian soil".[66] The issue of the relations between Orce and Sabadell and Andalusia and Catalonia regarding ownership of the bone will be addressed later, but it is worth noting that in this first brief remark, the journalist already compared the Orce bone with the *Dama de Baza* (*Lady of Baza*). In 1971, an ancient Iberian sculpture was found in Baza, a town very near Orce. After some controversy, the *Dama de Baza* was taken to Madrid, supposedly merely for its scientific study to be later returned to Baza for exhibition (see map 1). Finally, the sculpture remained in Madrid and was exhibited, and still is, in the *Museo Arqueológico Nacional*.[67] This journalist did not want history to repeat itself with the Orce bone fragment.[68] Later, even the mayor of Orce, José María González Galera, said that what had happened with the *Dama de Baza* would not happen again with the Orce Man.[69]

On 11 June 1983, a press conference was held in the assembly hall of the *Diputación de Granada*. The three discoverers attended the conference together with the president of the *Diputació de Barcelona*, the president of the *Junta de Andalucía*, the president of the *Diputación de Granada*, and some senior Andalusian scientists. During this press conference, politicians from the *Diputació* and the *Junta* signed an institutional scientific collaboration agreement that included new funding for the discoverers, excavation permits, and collaboration between both institutions.[70] As one journalist put it in one of the first articles dealing with the discovery, this collaboration meant the "first interdisciplinary team experience between institutions from different autonomous communities".[71]

The bone enjoyed an enthusiastic reception in Spanish newspapers which emphasised that "all media have covered the discovery" and announced it as the "finding of the century" and as a "scientific revolution". According to journalists, the discovery "defied current theories" about the peopling of Eurasia.[72] Newspapers also highlighted the fact that the presentation of the bone had been delayed for almost six months since its discovery in order to verify its

66 "pueda mantenerlo en tierras andaluzas", Montero, *ABC* 1983.

67 Contreras 2006.

68 Montero, *ABC* 1983.

69 Castro, *El País* 1983.

70 ADB: Caixa 4383. Exp. 3, *Instiut Paleontologia Miquel Crusafont*. See also: Fernández, *El Ideal de Granada* 1983.

71 "una primera experiencia de equipo interdisciplinar entre organismos de comunidades autónomas diferentes", De Semir, *La Vanguardia* 1983.

72 "todos los medios han cubierto el descubrimiento", "hallazgo del siglo", "revolución científica", "trastoca las actuales teorías", Aguilar, *El País* 1983; Redacción, *Mediterraneo* 1983; Moreno, *El País* 1983; De Semir, *La Vanguardia* 1983.

actual relevance.[73] There was also some criticism, though it never questioned the actual classification of the bone as hominid, but instead found fault with the overemphasis on the significance and antiquity of the bone.[74] Several articles discussed the economic situation of the scientific research, the *Institut*'s "present lack of scientific staff", and the discoverers' "confidence that, given the importance of the finding, the necessary means to move forward will be achieved."[75] The discovery even had some international echo. British newspapers like *The Guardian* and *The Times* highlighted the statements of Catalan prehistorian Eduard Ripoll (then director of the *Museo Arqueológico Nacional* in Madrid and former director of the *Museu d'Arqueologia de Catalunya*) that the discovery could be "one of the most important findings of recent years".[76] Another piece in *The Times* featured statements from Gibert, who highlighted how Peter Andrews had helped them "realise the broader significance of the discovery" and said that in the future they were planning to date the bone using palaeomagnetism in *Oxford University*.[77]

The 'Orce Man' nickname was also established during these first few weeks after the press conference. At first, journalists used either 'Hombre de Venta Micena' or 'Hombre de Orce', but by the end of June 1983, 'Hombre de Orce' ('Orce Man') was used everywhere, even among scientists.[78]

Privately, the bone also enjoyed an enthusiastic reception. After the press conference, Crusafont received a letter from Henry de Lumley that unfortunately has not been preserved. Luckily, he kept a copy of his own response to de Lumley and in it, the latter asked the Catalan palaeontologist to examine the fragment and suggested that they publish the finding together in a book.[79] Jaume Truyols, a leading Spanish palaeontologist and an early disciple of Crusafont's, also sent a letter to his former teacher in which he congratulated Gibert and his team on the discovery, stating that he had read about it in the newspapers,

73 Aguilar, *El País* 1983, and Lyon, *El País* 1983a.

74 For instance: Redacción, *ABC* 1983a.

75 "el actual déficit de personal científico", "confianza de que, dada la trascendencia del hallazgo, se obtendrán los medios necesarios para salir adelante." De Semir, *La Vanguardia* 1983.

76 Bill Clemyn-Jones, *The Guardian* 1983 and Debelius, *The Times* 1983.

77 Wigg, *The Times* 1983.

78 For an example of an early use of 'El Hombre de Orce' by scientists themselves see: Agustí/Gibert/Moyà-Solà, 1983d.

79 AMC-ICPMC: Letter from Crusafont to de Lumley, 3 August 1983. Original in French: "Quand a la collaboration a l'oeuvre sur lez Anteneanderthaliensis nous pouvons nous compromotre avec un article survle nouveau hom de Orce qui irait signe par moi et mes collaborateurs."

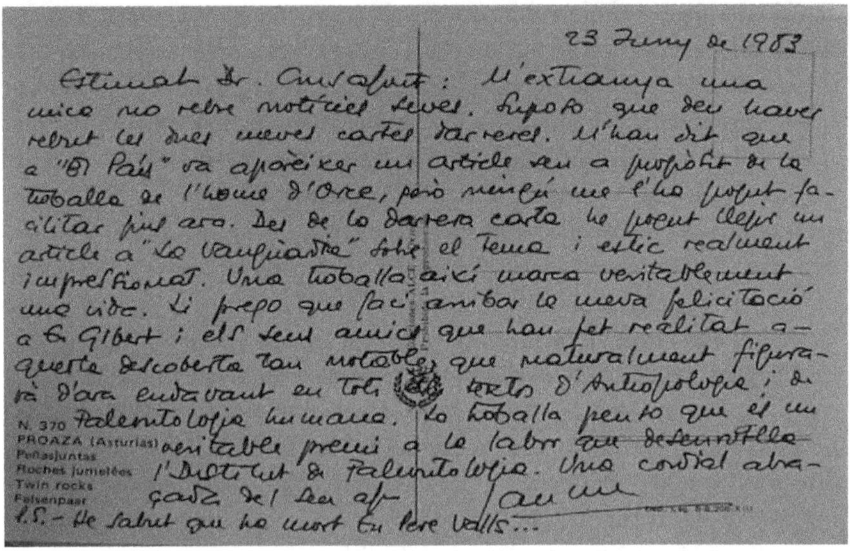

23 Juny de 1983

Estimat Dr. Crusafont: M'extranya una mica no rebre notícies seves. Suposo que deu haver rebut les dues meves cartes darreres. M'han dit que a "El País" va aparèixer un article seu a propòsit de la troballa de l'home d'Orce, però ningú me l'ha pogut facilitar fins ara. Des de la darrera carta he pogut llegir un article a "La Vanguardia" sobre el tema i estic realment impressionat. Una troballa així marca veritablement una vida. Li prego que faci arribar la meva felicitació a en Gibert i els seus amics que han fet realitat aquesta descoberta tan notable, que naturalment figurarà d'ara endavant en tots textos d'Antropologia i de Paleontologia humana. La troballa penso que és un veritable premi a la labor que desenvolupa l'Institut de Paleontologia. Una cordial abraçada del seu af. Jaume

P.S. - He sabut que ha mort en Pere Valls...

N. 370 PROAZA (Asturias) Peñasjuntas Roches jumelées Twin rocks Felsenpaar

FIGURE 4 Letter from Truyols to Crusafont congratulating Gibert and his team for the discovery of the Orce Man
FROM AMC – ICPMC

and finally adding that this "amazing" discovery would from then on appear in all "anthropology and human palaeontology textbooks" (see figure 4).[80] In his letter, Truyols did not present any kind of critique or disagreement with the scientific classification of the bone, with the methodology used by the three discoverers or with the public presentation of the new 'First European'.

During the following days, several newspapers accounts highlighted again how the Orce hominid demonstrated that Andalusians were our ancestors. Headlines like 'We are all Andalusians' or 'Adam's children were Andalusians' were common and some journalists even used the discovery to criticise exclusionary practices against Andalusians in Catalonia.[81] During the few preceding years (from 1975 to 1980), Andalusia witnessed a process of 'construction' of the so-called 'Andalusian identity' which would allow for the justification of a level of autonomous self-government similar to that of the 'historical nationalities', such as Catalonia or the Basque Country.[82] The Orce hominid was then presented as proof that Andalusia also had its own particularities. The director

80 AMC-ICPMC: Letter from Jaume Truyols to Miquel Crusafont, 25 June 1983.

81 "Todos somos andaluces", "Los hijos de Adán eran andaluces", Pernau, *El Periódico* 1983; Ramos, *Diario de Granada* 1983; Lyon, *El País* 1983a.

82 Soto et al. 2015.

of the *Diario de Granada* wondered: "Who could now claim that men from the south, the sons of the Orce hominid, are a people without history, without identity?". Later, this editorial presented the difficult economic situation experienced in poor regions in Andalusia, like the area where Orce is located.[83] Apparently, this process of identity construction had little to do with ethnicity and was more so grounded in left-wing social mobilisation that understood self-government as the solution to the problems of backwardness in Andalusia, compared to the rest of Spain.[84] The presentation of the Orce bone in the press was thus part of a wider process of 'construction' of the 'Andalusian identity', linked always to the social problems of rural Andalusia. The Orce Man was then part of the 'invented' Andalusian tradition. It was part of the creation of an historical continuity from the 'First Europeans' to the present Andalusians.[85] The press's concerns regarding possession of the bone also indicate how Andalusians saw control over their cultural heritage as a solution to their backwardness. This palaeoanthropological research was, for the Orce population, a way to attract tourism to the area and could therefore create significant economic activity for the people of the town.

After expert statements and the press conference, the bone was accepted as the 'First European', without taking into account the scientists' earlier doubts. These doubts, on both the anatomical characteristics of the bone and especially its dating, had completely vanished and were not presented to the media at all. Although newspapers echoed some external criticism on the dating, the general absence of these earlier doubts from the media meant that they also disappeared among scientists, as Truyols' and de Lumley's letters reveal.

∴

Two days after the press conference, on 13 June, the Orce Man was presented in the little town of Orce on what was described as the town's "happiest day". A small museum to exhibit the Orce bone and other remains form Orce's sites was inaugurated. Again, the *Junta de Andalucía*'s head of culture and the president of the *Diputació de Barcelona* chaired the event and the local band joined the entourage, "without having enough time to learn the Andalusian anthem".[86]

83 "¿Quién podrá negar a partir de ahora que los hombres del Sur, los hijos del Homínido de Orce, pertenecen a un pueblo sin historia, sin identidad?", Ramos, *Diario de Granada* 1983.

84 Soto et al. 2015.

85 Hobsbawm/Ranger 1992, 7. See a similar example with the famous Pekin Man in Leibold 2006 and Schmalzer 2008; and with the Brazilian remains of 'Luzia' Gaspar/Santos 2009.

86 "dia más feliz", "sin haber tenido tiempo suficiente de aprender el himno de Andalucía", Hernández, *Diario de Granada* 1983. See also Moreno 2014, 184 and 185.

After this presentation, the "*Dama de Baza* syndrome", as one journalist put it, resurfaced. The fragment had to be transferred from Orce to Sabadell to continue the scientific research, yet newspapers pointed out that the *Universidad de Granada* had enough good scientists to study the bone there.[87] People from Orce even blocked the scientists' access to the bone until a commitment was made regarding its return to Orce.[88] It seems that the Catalan scientists alleviated fears by signing an agreement with the *Junta* that "the bone will return to Orce within a year" and, of course, by declaring this in the media.[89] Some days later, the press announced that the Orce bone was in a safe in Sabadell since the next step in its study (preparation of the inner part) could not begin until the end of the summer period. This article also highlighted that the fragment was taken to the safe for security reasons because Gibert claimed to have "lost sleep" over the "mass panic" that had been created around the bone.[90]

On 24 June, Gibert stated in *El Periódico* that the excavations at Orce were "in danger" due to the "lack of funds" and that, so far, only "volunteering" had allowed the excavations to go ahead. The article also highlighted the threat of the site's "colonisation", highlighting that if foreign scientists came they would have a lot more resources than the Catalan scientists.[91] A few days later, on 28 June, the bone fragment was presented to the press in the *Institut de Paleontologia de Sabadell* with the presence of Antoni Dalmau, president of the *Diputació*, Jordi Labòria, head of the Department of Culture of the *Diputació*, and Eduard Porta, museologist and head of the Museums Department of the *Diputació*, together with Gibert, Agustí, Moyà-Solà, and the local secondary school students that went to excavate the previous summer (figure 5). The presentation was reported in local and national newspapers.[92] According to the press, Dalmau "pledged" to help with the development of the *Institut* and its research, though admitting that they had not yet defined how this would happen.[93] Dalmau also stated that the *Diputació* could not afford "excessively high costs", thus the help of the Spanish Ministry would be necessary.[94] The article in the local *Diari de*

---

87 "Síndrome de la Dama de Baza", Redacción, *ABC* 1983b; Redacción, *ABC* 1983c; Redacción, *La Vanguardia* 1983b.

88 Castro, *El País* 1983b.

89 "el fragmento volverá a Andalucía dentro de un año", EFE, *ABC* 1983 and Redacción, *El País* 1984a.

90 "una especie de psicosis creada en torno a ellos y que, según nos ha manifestado, 'me ha quitado el sueño' ", M. S., *ABC* 1983.

91 "Peligran las excavaciones", "por falta de medios", "voluntarismo", "colonización", Ramentol, *El Periódico* 1983b.

92 J.A., *Diari de Sabadell* 1983a and Barata, *El Periódico* 1983.

93 "se comprometió", J.A., *Diari de Sabadell* 1983a.

94 Barata, *El Periódico* 1983.

FIGURE 5 From left to right: Jordi Agustí, Josep Gibert, Antoni Dalmau, Jordi Labòria, Salvador Moyà-Solà during the press conference in Sabadell
FROM REDACCIÓN, *NOTICIERO UNIVERSAL* 1983

Sabadell also admitted that the finding had changed the way that the *Diputació* funded the *Institut*. For this local newspaper, the discovery turned Sabadell into the "Mecca of Palaeontology" and transformed the *Institut* into a world-class research centre at the same level as the Parisian *Musée de l'Homme* or the National Museum of Kenya.[95] Meanwhile, *El Periódico* printed that "The *Diputació* will help the Orce researchers" and highlighted Gibert's statement that "this help will be enough to continue the research", in a clear reference to Gibert's previous appearance in the same newspaper just four days earlier.[96]

In mid-July, the press also announced the start of the excavation expedition in Orce that summer. The bone went back to Andalusia with the scientists to be exhibited in Orce's small museum.[97] Since the press conference and until the excavation period, news about the discovery appeared very often, at least once a week. The 'Hombre de Orce' and Gibert, as its most visible discoverer, were becoming well-known characters in the Spanish public sphere.

Staff from the *Museo Arqueológico de Granada* and people from the *Universidad de Granada* joined the excavation expedition that summer of 1983. Isidro

95 J.A., *Diari de Sabadell* 1983b.
96 "La Diputación ayudará a los investigadores de Orce", "con estas ayudas será suficiente para poder continuar con las investigaciones", Gibert in Barata, *El Periódico* 1983.
97 Redacción, *ABC* 1983d. and Redacción, *El Periódico* 1983a.

Toro Moyano, an archaeologist from the *Museo* was among the additions to the group. Pascual Rivas, a palaeontologist from the *Universidad de Granada*, also started a scientific project linked to the Orce research. These two characters were prominent members of the PSOE, the political party that governed Andalusia and Spain at that time, and would become important characters in the Orce story. Gibert, Moyà-Solà, Agustí, and the secondary school students from Sabadell formed the rest of the excavation team, again. The excavations were held until the end of August.[98]

The press's use of the idea of the 'First European' being Andalusian, the emphasis on details like the failure of Orce's local band to play the Andalusian anthem, the fear of losing the discovery, and the inclusion of Andalusian scientists in the research show how, as discussed earlier, the making of the 'Andalusian identity' was at stake at the time and the renowned scientific discovery could not escape it. Likewise, the way that the bone was moved around, how it was sometimes showed in public presentations, and sometimes keep private, show how important was to 'control' the material scientific object, even beyond public and private discourses. The three discoverers controlled the object, so they controlled the public and scientific discourses build around it, and therefore the knowledge build around it. To have the bone, the scientific object, means, in successful scientific stories, to control what is said or can be said about it.

As indicated previously, on 3 August 1983, Crusafont sent a letter to Henry de Lumley answering a previous letter from the latter in which he presumably asked about the Orce Man. Crusafont explained the intended excavation period in Venta Micena to de Lumley.[99] According to Gibert, the very well-connected French prehistorian, Eduard Ripoll, phoned him to announce that the de Lumley couple wanted to see the bone and visit Orce and the excavation site.[100] In Orce, Marie-Antoinette, expert in anatomy, examined the bone and confirmed that it was hominid. Together, the researchers toasted to the Orce Man, the 'First European', in the Mari Cruz bar in Orce's town square. The de Lumleys invited the discoverers to study the bone in Marseille, where Marie-Antoinette worked. Henry de Lumley was also quoted in the press praising the Orce excavations.[101] The de Lumleys' approval was seen by the discoverers as the necessary international support for the discovery.[102] It also highlighted,

98 Gibert 2004, 39.

99 AMC-ICPMC: Letter from Crusafont to de Lumley, 3 August 1983.

100 Gibert 2004, 40.

101 *Ibid.* and González, *La Vanguardia* 1984.

102 Gibert 2004, 40 and Interview with Agustí 2012.

and again given the international context of the 'hunt for the First European', how at that time the work of Gibert, Agustí, and Moyà-Solà was not a marginal 'Spanish thing', but cutting-edge research in a hot topic that attracted international interest and was followed by leading scientists abroad, at least in France.

1.3 The 'Spanish Olduvai' and the Discoverers' Reward

That same day, after having lunch with the de Lumleys, the discoverers were informed that Crusafont had died.[103] Gibert went back to Barcelona and met the politicians from the *Diputació*. Later, they decided that Gibert would become the acting director of the *Institut*. Moyà-Solà and Agustí also gained permanent positions as researchers in the *Institut*, which changed its name to *Institut de Paleontologia Miquel Crusafont*, and multiplied its budget from the *Diputació* fivefold, from 13 to 60 million pesetas.[104] Together with the Orce discoverers, Lourdes Casanova, expert Tertiary mammals and dinosaurs became permanent member of the Institut. Apparently, funding from the Spanish Ministry of Science and Technology was also about to be given to a project conducted by the Sabadell researchers.[105] In less than a year, Gibert had turned himself from a secondary school teacher with a part-time, often voluntary, job in palaeontology into the head of the new *Institut de Paleontologia Miquel Crusafont*, which became the main palaeontological institute in Catalonia and one of the leading institutes in Spain, with a new budget, four new full-time researchers, and more funding coming from the Spanish central government. It could even be argued that the discovery of the Orce Man and its promotion through Spanish media 'saved' the *Institut* from probable disappearance after Crusafont's death, or at least accelerated its revival.[106] The *Diputació* now had a powerful reason to maintain and increase the money spent on the *Institut*: the famous and amazing 'First European' and its discoverers.

It seemed that the popularisation of the finding had paid off and the discoverers' efforts to stress the need for financial help had led to better positions and funding. As Gibert, Agustí, and Moyà-Solà put it in a series of full-colour newspaper articles that described the characteristics of the bone and the site, the scientific conditions in which it was found, and the Orce Man's very important

103 Gibert 2004, 40 and Redacción, *La Vanguardia* 1983c.

104 Gibert 2004, 41.

105 Castro, *El País* 1983 and Redacción, *Noticiero Universal* 1983.

106 Interview with Agustí 2012; Interview with Moyà-Solà 2012; and Interview with Roger Marcet 2013.

position in the 'humanisation process': "with the necessary institutional support [...] we hope to fulfil our responsibility".[107] The researchers presented their work in the press as prominent and crucial, but at the same time as precarious and in need of funding from institutions. Popularisation then became not only a way to echo an important scientific discovery, but also a way to attract politicians' attention in a successful bid to secure funding for future research.

In December of that year, *Arrel*, the *Dipuatació de Barcelona*'s magazine, published a series of articles on the *Institut de Paleontologia*, the Venta Micena site, and the Orce Man, the first two anonymously signed and the last one with Gibert, Agustí, and Moyà-Solà as authors.[108] According to the anonymous articles, the *Institut* was "a little-known scientific institution", but after the discovery and public announcement of the Orce Man the three discoverers became "known to everyone".[109] These articles also emphasised how the research had been conducted up until then with monk-like austerity but that now funding would come from all the institutions: the *Diputació de Barcelona*, the *Junta de Andalucía*, and also the central government.[110]

After the summer excavations, Gibert went for the first time to Marie-Antoinette de Lumley's laboratory in Marseille.[111] There, they analysed the bone in detail and compared it to other hominid craniums. It seemed that without any doubt, VM-0 belonged to a hominid. Marie-Antoinette and Gibert agreed to meet again when the inner part of the bone was visible. The de Lumleys appeared to be reinforcing their position as the international experts that 'supervised' the Orce Man's scientific dissemination. The new 'First European' remains were then under their control. In October, the bone was taken to the *Museu d'Arqueologia de Catalunya*, which was also managed by the *Diputació de Barcelona*, where a team of technicians under Campillo's supervision began to clean the inner part.[112]

The fragment's transfer from Orce to Sabadell and back to Orce was followed almost daily in Spanish newspapers. The notion of a poor periphery in the countryside (the little town of Orce) and an apparently dominant centre (the prestigious *Institut de Paleontologia* and its researchers) emerges from

107 "con el apoyo institucional necesario [...] esperamos cumplir con nuestro cometido", Agustí/Gibert/Moyà-Solà, *La Vanguardia* 1983.

108 Anonymous 1983a.; Anonymous 1983b.; Agustí/Gibert/Moyà-Solà, 1983b.

109 "Una institució científica poc coneguda"; "coneguts per tothom", Anonymous 1983b, 69.

110 Anonymous 1983b, 70.

111 Gibert 2004, 44.

112 *Ibid.*, 45; Interview with Agustí 2012; Interview with Campillo 2011; Campillo 2002, 29–31.

the analysis of these newspaper pieces.[113] Once again, peripheries are economically poor but rich in resources (in this case the Venta Micena site) and therefore try to claim their 'treasures' before they are controlled by centres with more economic potential. Meanwhile, scientists wielded their right to the scientific study of the piece and complained about the possibility of another 'centre': the rich foreign scientists that could potentially gain control of these resources. Yet, these centre-periphery discourses hide the different parties' interests in benefiting from the situation. The Orce Town Council demanded that the valuable remains be returned to their 'possession', while the town was widely advertised in national media. The possible tourism potential brought about by the discovery in the little town of Orce surfaced during the first days after the presentation. It seemed that the town's shopkeepers very soon took advantage of the discovery by selling souvenirs of it.[114] At the same time, Sabadell, which is in the periphery of Barcelona both geographically and in terms of research, vindicated the *Institut* as a possible world research centre in the area of palaeontology. A press presentation of the discovery in Sabadell was necessary, as it also was in Orce. The *Institut* gained promotion and prestige in the public sphere while politicians also gained publicity in the town.

In addition, scientists used this geographic situation in a two-sided discourse regarding the foreign presence in their research. This dichotomy will be discussed in depth later but already appeared in this first phase of the Orce story. On the one hand, foreign experts were presented as necessary for confirmation of the finding's validity. The de Lumleys' interest in publishing the bone under their supervision was also seen as international validation. On the other hand, the discoverers, and especially Gibert, publicly referred to a possible 'foreign colonisation' of sites and remains as part of their effort to get funding from the *Diputació*.

Moreover, the de Lumleys' themselves (perhaps the well-funded foreign scientists Gibert was concerned about), coming from the most prestigious institutions in Paris, were necessary for the researchers as an international blessing for the finding, but they also used their power to control the way that the new 'First European' would be published and disseminated.[115] In addition, France

113 For more on the centre-periphery problem see: Gavroglu et al. 2008; and with a special focus on science and its 'publics' see Papanelopoulou/Nieto-Galan/Perdiguero (eds.) 2009.

114 Pedregosa 1983.

115 For another comment on this foreign colonisation see Anonymous 1983c, 39. You can find a similar case in Atapuerca with the very same de Lumley in Hochadel 2013b, 61–72.

and the French scientists represented for the Spanish researchers, and later for the Spanish general public, the democratic state that was under construction at the time in Spain. France was what Spain somehow wanted to be in terms of democracy.

Nevertheless, scientists were not the only ones that benefited from this complicated situation. The *Diputació de Barcelona*, though its work covered the entire Barcelona province, was located in the city of Barcelona itself. Likewise, Andalusia is governed from Seville, where the Andalusian parliament is. Both institutions, despite not being physically located in Orce or Sabadell, used the discovery for their political promotion in those areas. Both institutions ensured their control over discoverers, the scientific institution, the bone, its popular stance and the site in order to develop their own promotion as new (re)formed political institutions. Of course, the great discovery was also used for promotion beyond those areas. For instance, the *Junta de Andalucía* presented a reproduction of the Orce Man in an 'Expo-Train' that travelled all around Spain promoting Andalusian culture and tourism.[116] The *Diputació* publicly announced its financial support of the researchers several times in newspapers and promoted the discovery in its own magazine.[117] And all of this happened in a very concrete political context: the new democratic state and, within it, the consolidation of the newly formed institutions, such as the *Junta* and the Autonomous Community of Andalusia, and the redefinition of old structures, such as the *Diputació de Barcelona*.[118] At the same time, the way this promotion worked was closely related to the political parties in office in both institutions.

All of these complicated geographical relations between Sabadell, Barcelona, Orce, Granada, Seville, and Paris (see map 1) show how the simple concept of a dominant centre and a weaker periphery (both economically and in knowledge making) is not enough to explain the geographical interactions between institutions, researchers, and town councils when a scientific discovery like this takes place. A more complex analysis taking into account needs, interests, politics, and prestige is necessary. From this kind of analysis it will be much easier to go beyond the 'easy brands' of centre and periphery and focus this study on the actual geographical location of institutions, scientists, and, in our case, also sites, in order to show how this influences the

116 Oliveras, *El Periódico* 1983 and Redacción, *El País* 1984b.
117 Anonymous 1983a.; Anonymous 1983b.; Agustí/Gibert/Moyà-Solà 1983b.
118 Jordi Agustí himself acknowledges that the political situation influenced the way that the bone was presented to the media, Interview with Agustí 2012.

way that science develops and circulates among and between the different actors.

All this is in fact related to the concept of 'geographies of knowledge'.[119] Trying to follow the importance of place in shaping science, recently researchers have overcome the traditional view that science is mainly developed in the great the 'centres' of knowledge, metropolises, universities, and research institutions, and have shown how often the contribution of the supposed 'peripheries' is significant and how even when a new scientific concept arrives from the 'centre' it is 'appropriated' by peripheral actors who adapt and shape it depending on the local conditions. This new research has led to the questioning of the notions of 'centre' and 'periphery'.[120] In this regard, the volume *Popularizing Science and Technology in the European Periphery, 1800–2000* has been of paramount importance for relating these geographical notions to the publics of science.[121]

In November 1983, the Orce Man appeared again in the newspapers. At the launch of a geology exhibition in the *Universidad de Granada*, Gibert stated to the press that the summer excavations had revealed that Venta Micena not only held the 'First European' but was also 'the oldest human settlement in Europe and Asia'. At the end of 1983 and the beginning of 1984, this claim appeared repeatedly in the Spanish press.[122] Animal bones found in Venta Micena had a distribution pattern that, according to the researchers, indicated human agency, which led them to think that the site was not a casual accumulation of bones or a den of scavengers, but a hominid settlement. With this claim, a 10 cm cranial bone fragment was transformed into a one-million-year-old human settlement in Europe. Agustí and Moyà-Solà even claimed that more 'Orce Men' were about to be discovered.[123] These claims did not appear in the first academic article that discussed the discovery. Therefore, the press became a channel for presenting brand new scientific research. The Orce man was then

119 See for instance: Livingstone 2003 and Naylor 2005.

120 For a comprehensive account of the research carried out on centres and peripheries, and of the concepts themselves, see Gavroglu et al. 2008.

121 Papanelopoulou/Nieto-Galan/Perdiguero 2009.

122 "el asentamiento humano más antiguo de Europa y Asia", Gómez, *ABC* 1983. See for instance: Manso, *ABC* 1984; Paredes/Moix, *La Vanguardia* 1984; González, *La Vanguardia* 1984.

123 J.A., *Diari de Sabadell* 1984.

a success, so there was no problem to present this kind of claim in the press. In addition, journalists were eager for news regarding the Orce discovery and the researchers took advantage of this situation, placing their claims and themselves in the public sphere.

During this time, mainly Gibert, but not only him, also put forward the hypothesis that hominids came to the Iberian Peninsula across the Strait of Gibraltar.[124] The August 1983 issue of the Spanish magazine *Revista de Arqueología* featured some articles that dealt with human origins and how humans spread to Europe (figure 6). Among them, Gibert, Agustí, and Moyà-Solà published an article that described VM-0 and considered two possibilities for hominid arrival to Europe: the Levantine Corridor hypothesis, the most accepted view at that time, and the Strait of Gibraltar crossing, which, according to the researchers, could not be disregarded since Orce was located in southern Spain. Some newspapers widely and clearly reproduced this claim highlighting the 'amazing adventure' of the Orce Man crossing the Strait.[125] As in the case of the Venta Micena human settlement claim, scientists used the press to introduce research findings and hypotheses (which had not been presented in any scientific forum) to the public sphere. This strategy allowed the researchers to maintain public interest in the discovery while also presenting new claims to their peers through the general media. The *Revista de Arqueología* is a good example of a specialist yet at the same time popular magazine that aims to reach both audiences and forms part of the "spaces in which scientists can creatively explore the wider implications of their claims."[126] When these claims are controversial, or could become controversial, like the Strait of Gibraltar crossing, scientific popularisation can become a test field to prove and strengthen them before, or at the same time as, their presentation in scientific forums.[127] By repeating over and over again in the media that the Venta Micena site was the oldest human settlement in Europe, the researchers tried to make this new knowledge more visible and therefore harder for the scientific community to contradict. This way, newspapers and magazines become yet another place in which to present scientific knowledge to society and, more crucially, also to colleagues.

124 Gibert would maintain this claim throughout his life. See for instance: Gibert/Gibert/Iglesias 2003.

125 Agustí/Gibert/Moyà-Solà 1983c and Ramentol, *El Periódico* 1983a.

126 Goulden 2013.

127 For further analysis of this point see Bucchi 1996, 375–394. In palaeoanthropology, a good example of this phenomenon could be Goulden's 2013 review of the presentation of the world famous *Homo floresiensis*.

FIGURE 6 Front cover of the *Revista de Arqueología* with Orce's article
REPRODUCED THANKS TO ZUGARTO EDICIONES

Moreover, the *Revista de Arqueología* article is also an example of how the three discoverers made a remarkable popularisation effort during the first year after the press conference. They used different media in their legitimising discourse around the Orce Man: they gave lectures to a wide range of 'publics' (Gibert acknowledged to the press that they "fill conference halls"), appeared on radio and television, and wrote their own newspaper and magazine articles.[128] As *El Periódico* put it, the discoverers were, by the beginning of 1984, a "famous team of palaeontologists".[129]

And that was not all. In October 1983, just four months after the public presentation of the discovery, Josep Gibert (representing the team) signed a publishing contract with the Catalan publishing house Empúries for a popularisation book provisionally entitled *The origins of the human settlement in Europe: The Orce Man*. This publishing house had just been founded on 1983 and their first books did not appear until March 1984, so the Orce Man book was supposed to be one of its first releases.[130] The publisher's representative that signed the contract with Gibert was Xavier Folch i Recasens, one of the founders of the house and a member of the newly formed Catalan parliament, representing the communist party PSUC, of which Gibert was also a member.[131] The contract stipulated that the authors would receive 300,000 pesetas in royalties in advance, with 100,000 of that amount when the contract was signed.[132] Following historian Oliver Hochadel, it is necessary to highlight here that popular science books are much more than just a way to spread scientific knowledge among lay people; they are a place where researchers present their own stories and their research in a way that it could not be presented in a scientific article and to a wider range of audiences, including researchers in their own field.[133] It is also worth noting that the provisional title of the book, as well as the researchers' lectures titles, used the popular name 'Orce Man', by then identifiable and used by everyone in all types of contexts.

128 "llenamos salas de conferencias", R.F., *El Noticiero Universal* 1983. For talks see for instance: Gibert 'L'Home d'Orce: L'habitant més antic d'Europa' in the *Museu de la Ciència de Barcelona* from Redacción, *La Vanguardia* 1984. By Gibert, Agustí, and Moyà-Solà in the *Institut d'Estudis Catalans*, 'L'Home d'Orce i el seu entorn cultural vist a través de les indústries trobades recentment a Venta Micena (Granada)' from Redacción, *La Vanguardia* 1984b. And by Gibert, Agustí, Moyà-Solà, Eudald Carbonell, Domènec Campillo, Jordi Sabater-Pi and J. F. De Villalta in the *Museu Arqueològic de Barcelona*, 'L'Home d'Orce. El Primer Habitant d'Eurasia' from Campillo 2002, 29 and Redacción, *El Periòdico* 1983b. Gibert and Agustí on the Radio: Redacción, *ABC* 1983e. For Magazines: Chamorro 1983; Pedregosa 1983; Redacción Algo 1984; Anonymous 1983c. And Gibert on TV: *Comarques*, TVE from Redacción, *El Periódico* 1983c; UHF 1983, La Clave; TVE 1983, La Tarde. To compare the articles appeared in this period with other periods of the controversy see Annex 2.

129 Redacción, *El Periódico* 1984a.

130 'Provisionalmente titulado: Los Orígenes del Poblamiento Humano en Europa: El Hombre de Orce', AJG-ICPMC: Contracte d'Edició (publishing contract), Barcelona, 7 October 1983, and Piñol, *La Vanguardia* 1984.

131 M.S. *Vilaweb* 2008 and Anonymous 'Xavier Folch i Recasens'.

132 AJG-ICPMC: Contracte d'Edició (publishing contract), Barcelona, 7 October 1983.

133 Hochadel 2013b, 252.

To understand how the Catalan scientists acted, it is again necessary to consider the wider international context of popularisation of palaeoanthropology. In the late 1970s and early 1980s, it seemed that the popularisation of the findings and theories of Donald Johanson and Richard Leakey (and the controversy between them) was having an incredible impact in Spain.[134] For instance, as Agustí acknowledges, Johanson's book, *Lucy: the beginning of mankind* (translated into Spanish just a year before the Orce Man discovery), "was a hit" among the Spanish scientific community.[135] It also seems that the Orce Man discoverers used some of Leakey's and his team's representations of *Homo habilis* in their popularisation of the Orce Man.[136]

In a way, with the new budget given to the *Institut* and with all these popularisation efforts that subjected the Orce Man finding to different kinds of media coverage, can be argued that the discoverers were somehow 'applying' for Miquel Crusafont's position within Spanish palaeoanthropological community. This application was addressed not only to the scientific sphere, but also to the public sphere, where Crusafont had been a well-known figure for many years.[137] It seems they were trying to take Crusafont's 'niche' public position as the Spanish/Catalan palaeontology-archaeology expert for the media. When Crusafont died, Gibert wrote an article in a Catalan newspaper in which he reviewed his experience working with Crusafont and finished stating that Crusafont's scientific work and the *Institut* as a research institution formed the 'inheritance' that he had left to his collaborators.[138] At the same time, the Orce Man and the discoverers' popularisation efforts helped to consolidate palaeoanthropology as a discipline in Spain after Franco's dictatorship. The *Institut de Paleontologia*'s recovery was a tangible representation of how professional palaeoanthropology and palaeontology had been strengthened.

However, Orce was not alone. At the same time as the Orce Man was presented, several hominid bones were about to be found in the Atapuerca Mountains. Since the early 1990s, the Atapuerca research project had developed a successful relationship with the Spanish media that led to what historian

134 Lewin 1997. In Spain, see, for instance, Guardia, *La Vanguardia* 1979. For Richard Leakey see Leakey 1985. For Johanson see Johanson/Maitland 1982.

135 Interview with Agustí 2012; Hochadel 2013c, and Hochadel 2013b, 216, also talks about the impact of this book in Spain; Redacción, *La Vanguardia* 1982; Gibert also cited this work in his 2004 popular science book, Gibert 2004, 445.

136 According to Molina 1998, they used a reconstruction that Leakey made to illustrate the anatomical characteristics of the *Homo habilis*, with which they made a poster to disseminate their findings.

137 Florensa 2013; Florensa 2017 and Acosta 2013.

138 Gibert, *El Periódico* 1983.

Oliver Hochadel calls the "making of a Magic Mountain". For Hochadel, the success of Atapuerca in Spain is not merely a result of the amazing quantity and quality of its hominid remains, but can only be understood with a proper consideration of the popularisation 'industry' built around the site.[139] As I have already indicated, when science turns to media different actors and institutions get involved, and each with their own agenda.[140] In the Orce case, scientists, politicians, journalists, research institutes, newspapers, and town councils all had their own interests. It seemed that together all these actors were taking the firsts steps towards building a whole popularisation structure around the Orce Man and the Venta Micena site, in a similar way to what happened later in Atapuerca. With the Orce Man still a great success, they were trying to create an 'Orce Man industry' with museums, popular science books, media appearances ... Yet, the Orce case would not be a 'success story'. This brings to the surface a very interesting point. It shows how, despite the differences between subsequent accounts, unsuccessful stories could have exactly the same roots, the same processes, and the same strategies as those that are successful. Gibert, Agustí, and Moyà-Solà's research could have ended up being similar to what Hochadel describes in the Atapuerca case. Orce could thus have become not a 'Magic Mountain' but certainly a wonderful 'Spanish Olduvai'.[141]

The Orce discovery's media presence allowed easier visibility for other supposedly older sites that were already being researched or were being excavated at the same time. Two examples illustrate how Orce was, in Spain, a spark that led to the emergence of sites claiming to bear clues for understanding the 'First European' problem.

In the same *Revista de Arqueología* issue referred to earlier (figure 6), María Ángeles Querol and Manuel Santonja, two well-known Spanish archaeologists, wrote an article about the El Aculadero site, a possible early occupation of southern Spain near Cadiz (see map 1). This article also stated that the site suggested that entry into Europe through the Strait of Gibraltar happened in an early phase of human evolution.[142] The El Aculadero research shows that

---

139 Hochadel 2013b, 35–72.

140 Kjærgaard 2011a, 1,2.

141 Apparently, in his visit to Orce, de Lumley compared its sites with Olduvai, a comparison that has been repeated ever since in several media, including scientific articles. For instance: R. F., *El Noticiero Universal* 1983, 14; J. A., *Diari de Sabadell* 1983b; Zihlman/Lowenstein 1996.

142 Querol/Santonja 1983. In the scientific sphere, El Aculadero was claimed to be evidence of an arrival to Europe via Gibraltar, see: Thibault et al. 1977. See also Phillips 1980, 37.

Gibert was not alone in claiming the occupation of Europe through Gibraltar. Again, this show how for many researchers the Orce bone did not mean a great conceptual shift but in fact reinforced previous hypotheses about the early occupation of Europe.

In April 1984, Gibert, Moyà-Solà, and Agustí, among other scientists, such as Eudald Carbonell, participated in a press conference to announce the discovery of even older stone tools in the Casablanca site in Almenara, Castelló, in the Autonomous Community of Valencia (see map 1).[143] Once more, the Casablanca site became 'the oldest human occupation of Europe', even showing evidence of the presence of *Homo habilis*.[144] Although it had no hominid remains, and therefore less media attention, this new announcement had a double-sided interpretation in relation to Orce. On the one hand, it strengthened the discoverers' claims of an early occupation of southern Spain. On the other hand, however, for some it seemed to take significance away from the Orce findings. As one newspaper put it with tongue in cheek, "the oldest human settlement in Eurasia, which according to all the data was located a year ago in Venta Micena, Granada, is now in Castellón."[145]

The El Aculadero site and the Casablanca stone tools apparently later turned out not to be hominid-made or were not as old as claimed, but the way that these sites flourished in the public sphere shows once again how the claims made in Orce were not isolated and out of the blue but set in a prior national and international scientific scene.[146] Moreover, Orce was also actively followed by the members of this scene. The Orce Man popularisation efforts and the previously explained conditions that allowed its emergence in the Spanish public sphere also conditioned how these other sites appeared and were reported on in the press. In short, the discovery 'opened the door' for the press to other sites and researchers that without Orce's boom probably would not have had the same popular impact. To sum up, Orce sparked media interest in early European colonisation and transformed newspapers into a fertile ground for archaeological and palaeoanthropological news.

143 Prats, *La Vanguardia* 1984 and Redacción, *Mediterráneo* 1984. For more on this excavation see Gusi 2005a.

144 Andrés, *El País* 1984.

145 "'El asentamiento humano más antiguo de Eurasia, que según todos los datos se situaba, desde hace un año, en Venta Micena (Granada), está en 'Castellón'", Prats, *La Vanguardia* 1984. See also Gusi 2005a, 20–21.

146 For Almenara: Gusi 2005a, 22, 25, and 26. For el Aculadero: Lópz/Martínez 1997.

1.4 A Great Post-Franco Discovery and a Small but Troubling Crest

On 11 January 1984, *El Periódico* published a photograph of Antoni Dalmau, the president of the *Diputació de Barcelona*, showing the Orce Man bone fragment to Amadou M'Bow, general director of *UNESCO*.[147] Next to M'Bow, Pascual Maragall, mayor of Barcelona, also from the PSC-PSOE, and in the background a smiling Jordi Agustí. The photograph illustrated an article on the new findings in Venta Micena and their analysis. It highlighted that the discoverers' research "leads slowly but surely towards an authentic revolution in the old theories."[148]

This picture is yet another example that reveals how the political context, and not only the scientific importance of the finding, determined the way that the Orce Man was presented to the public. This context also shaped the way journalists treated and presented it, and the way it was received by politicians and the general public. Politicians pursued great achievements, journalists, who could write freely, were eager for big scoops, and people also craved major national events: "The windows were opened" after 40 years of dark dictatorship.[149] The Orce Man was a significant scientific discovery for this period, and maybe the first that received such media and public attention. As Gibert himself later stated "[The Orce Man] emerged during the years of journalistic sensationalism at the beginning of the democracy."[150] One could even argue that a different scientific discovery in some other discipline would have received similar treatment from scientists, politicians, journalists, and the public. The country wanted normality in democracy and therefore needed normality in science. In a way, people felt that Spain needed to catch up with other countries and overcome the backwardness caused by the dictatorship. What researchers wanted to ensure was that this media attention included its own discourse emphasising the 'lack of funding' for science, palaeontology and palaeoanthropology and for their own specific research.[151] The media thus became a place where resources and support for science were negotiated at a time when newly arrived politicians were actively deciding how and where these resources should be used. At the same time, what palaeoanthropology does easily is to involve a strong element of identity, closely linked in this case

147 Barata, *El Periódico* 1984.

148 "conduce, lenta pero progresivamente, hacia una auténtica revolución en las viejas teorías", *Ibid.*

149 "Les finestres es van obrir", Interview with Roger Marcet 2013.

150 "tuvo su origen en los años del sensacionalismo periodístico de principios de la democracia", Redacción, *El Mundo* 1999.

151 See a similar instance in González-Silva 2005 and González-Silva 2007.

to the political situation of post-Francoist Spain and the *autonomías*. The 'First European' was also the first achievement of the new Autonomous Community of Andalusia.

In addition, the way that scientists, politicians, and journalists dealt with the discovery has shown us how the Orce research could become a kind of scientific 'industry' for the Spanish public, a way to 'consume' science that could lead to a very successful and well-known site but that ended up being a completely different thing. For what comes next, it is also necessary to state that at least until April 1984, the Orce research and claims were not scientifically marginalised but were central among a group of well-positioned scientists that were doing research on hominid arrival to Europe. Therefore, the Orce example shows us how the scientific claims that these 'industries' entail could be successful or not depending on the successes of the entire enterprise. Scientific claims then do not exclusively depend on their own plausibility but also on the environment they are presented in, by whom, and how.

Finally, let us return to our story. At the end of 1983, a big international conference was announced to the press for the end of May 1984 in order to present the Orce Man in Granada to national and international scientists.[152] The conference would be organised by the *Junta de Andalucía* and would gather almost five hundred experts including (according to a newspaper) Richard Leakey.[153] By the end of April 1984, technicians from the *Museu d'Arqueologia* finished the cleaning process of the inner part of the bone. During this process, an unusual crest appeared.[154] The crest seemed to steer evidence away from hominid classification. For Campillo, everything remained the same, as this crest could also appear in humans.[155] But doubts surfaced again among the discoverers and they decided to bring the fragment to the de Lumleys, as they had previously agreed with Marie-Antoinette. They thus decided to rely again on the prestigious French discoverers. Just to get an idea of how the three young discoverers saw Herny de Lumley, it is worth noting that at that time he was known to them as 'Monsieur L'Empereur'.[156]

From here on in, the story changes radically. Let us now move from the Orce Man public discovery into the depths of the Orce Man public controversy.

152 González, *La Vanguardia* 1984; Redacción, *El Periódico* 1983d; R. F., *El Noticiero Universal* 1983; Redacción, *ABC* 1983f; *ABC*, 1983g.

153 Relaño, *El País* 1984b and R. F., *El Noticiero Universal* 1983.

154 Interview with Agustí 2012; Interview with Campillo 2011; Relaño, *El País* 1984b. See also Annex 1.

155 Interview with Campillo 2011 or Campillo 2002, 33.

156 Interview with Agustí 2012.

CHAPTER 2

Controversy (1984–1987)

Picture the scene: three young researchers arrive at their research centre just after a trip. One of them carries a box, which they do not pay much attention. Another takes the newspaper from his table. Suddenly, he lets out a shout: "That's impossible!" The other two run over to him. Something strange is happening: the scientific discovery of their lifetime is being questioned on the front page of a widely read national newspaper. After this front page appeared, a public controversy began.

When doubts about the validity of the hominid claims on VM-0, the Orce Man story was transformed from a successful public discovery into a long and arduous public controversy. Moreover, the early popularisation effort lead to a high level of attention to the later dispute. From 'successful public discovery' to 'arduous public controversy', what remains the same is the 'public' factor. Through this metamorphosis from discovery to controversy, 'scientific elites' (both at home and abroad) reinforced and protected their power and built a wall to try to keep out from them public controversy and disagreement. In the midst of all this, Josep Gibert i Clols emerged as the main character in the Orce controversy. He tried both to protect himself and to spread his word in an effort to maintain his professional and public position. At the same time, he began to 'construct' a characteristic discourse that would remain for several years. The Orce Man controversy spread through different forms of cultural expression and the media (especially newspapers) became a necessary part of the way that scientific knowledge claims sought to be validated and ultimately confirmed.

2.1 A Painful Trip to Paris: From Man to Donkey

Between 3 and 5 May 1984, Gibert again took the Orce bone to Marseille to show the cleaned inner part to Marie-Antoinette de Lumley.[1] She also thought that something was wrong with the inner features (figure 7). A ridge or crest on the bone did not match human anatomical characteristics. They compared it with other hominids and tried to place the bone in another part of the cranium.[2]

1 Interview with Agustí 2012.

2 Gibert 2004. For more detail on the anatomical features of the Orce bone see Annex 1.

 | DOI:10.1163/9789004431508_004

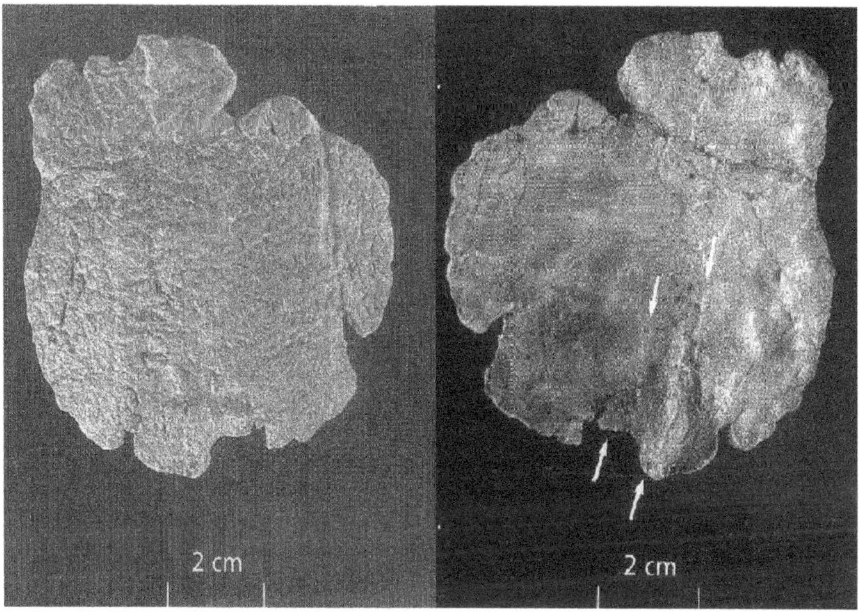

FIGURE 7 The cranial fragment after the inner part was cleaned. Left, external part; right, inner part
FROM CAMPILLO 2003, SHEETS 11 AND 12. PHOTO ORIOL CLAVELL

According to Gibert, during this visit and for the first time, Marie-Antoinette raised the possibility that VM-0, the famous Orce Man, could in fact be a young member of the horse and ass genus, *Equus*. Still, Gibert and de Lumley agreed to meet again some days later in Paris, where they could compare the fragment with other animal craniums that they did not have access to in Marseille or in Sabadell.[3] There, Henry de Lumley and the two other discoverers could also join them to discuss this new possibility. During this visit to Marseille, Gibert phoned Agustí, who was in Almenara, where a conference was taking place. While explaining the situation and the possibility raised by de Lumley, Gibert was very nervous, according to Agustí.[4]

Following Gibert's account, in the days between those two meetings in Marseille and Paris, Henry de Lumley phoned him at home. The French scientist urged the Catalan to give a press conference to announce that the Orce Man was indeed a young ass, which Gibert refused to do.[5]

3 Gibert 2004, 46.
4 Interview with Agustí 2012.
5 Gibert 2004, 46.

On 9 May, just before they went to Paris, Gibert, Agustí, and Moyà-Solà asked the *Junta de Andalucía* for a six-month postponement of the international conference that was scheduled for the end of the month.[6] Gibert later argued that they asked for a postponement because they had some problems with the Orce bone (mainly the strange crest) which could not be resolved in the time remaining before the conference. He also said that they did not want the conference to turn into a controversial encounter or even an "anti-Orce Man" event.[7] According to subsequent news pieces, the *Junta* sent out a very brief press release the next day to all media announcing the cancellation of the conference. Yet this announcement did not appear in any national newspaper.[8]

Around 10 May 1984, the three discoverers travelled by train to Paris. They went both to the *Institut de Paléontologie Humaine* and to the comparative anatomy laboratory of the *Muséum National d'Histoire Naturelle*, where they were able to compare the bone with several other animal specimens. Surrounded by thousands of 18th and 19th century animal bones from the *Muséum* collections, Marie-Antoinette de Lumley, expert in anatomy, concluded in front of the three discoverers that the Orce remain was not a hominid bone but rather a young ass. She opened the skull of a young onager (*Equus hemionus*), a present day Asiatic wild ass, to show them the inner part. According to Agustí, at that moment, he realised that they had been wrong.[9] For him, the Orce Man's inner anatomy, with the controversial crest, seemed to perfectly match the equine (figure 7). But Gibert held another position. He tried to convince the others that the bone needed more detailed study. He was not at all sure that they had been wrong. Already in the Parisian laboratory, Gibert did some research comparing the fragment to other skulls of colts and horses. According to him, Moyà-Solà helped taking photos and measurements while Agustí went back to the hotel.[10] For Gibert, more comparisons and research on the bone were required back in Barcelona before definitively excluding the hominid likelihood. Therefore, it seems that Agustí accepted the de Lumleys' verdict but Gibert still believed in the possibility that the fragment belonged to the genus *Homo*.[11] Apparently, Moyà-Solà remained in a middle position, helping Gibert, but apparently also with strong feelings that they were wrong. The de Lumleys again urged the discoverers to hold a press conference to admit their

6 Relaño, *El País* 1984b.
7 Redacción, *El País* 1984c.
8 Relaño, *El País* 1984b.
9 Interview with Agustí 2012 and Interview with Moyà-Solà 2012.
10 Gibert 2004, 48.
11 Interview with Agustí 2012.

mistake. Gibert, as the senior of the three, did not accept this and left the Paris laboratory in complete disagreement with the de Lumleys.[12]

This first 'private' phase of the Orce Man controversy serves as a good example for understanding some characteristics of what, applied to palaeoanthropological and prehistoric research in the 20th century, has been called 'scientific colonialism'.[13] On the one hand, the Spanish researchers needed the well-known French scientists' authority and expertise as well as the French museum's huge collection of skulls. It seemed to the Spaniards that with the French approval of the finding they would be able to enter international scientific circles, to publish, and to gain recognition. They relied on this international authority and material in order to try to solve the crest problem. But, on the other hand, when these experts contradicted the expected interpretation of the findings, Gibert rejected their opinion and decided to do more research back home.[14] Moreover, together with the de Lumleys' trip to Orce and their proposal for a joint publication, the way that Gibert and the discoverers went first to Marseille and afterwards to Paris shows how strong the de Lumleys' will to control the Orce Man research was. The French couple wanted to be part of the process of the scientific publication and popularisation of the finding. Yet, with Gibert's rejection of their authority, the de Lumleys would not have been able to keep the Orce remains under their control. Then, that same day, they made a crucial move in order to secure their powerful and dominant position in the European hominid 'hunt'. They did not want a 'First European' beyond their rule, so apparently they decided to tear the discovery apart. Moreover, they did so in the same way that the discovery itself had been 'constructed': publicly.

The next day, the three discoverers travelled back to Barcelona by train. When they arrived on the morning of 12 May 1984, bad news was awaiting them. *El País*'s front page headlined 'Serious indications that the Orce Man skull belongs to an ass'.[15] Inside, three articles filling two complete pages discussed the

12 Gibert 2004, 48.
13 Hochadel 2013b, 35–72.
14 This dichotomy has been summarised in Hochadel 2013b, 40–42, with a very concise slogan: 'Come and see what we have; Keep out! It's ours!' See another example in: Lanzarote 2013.
15 'Serios indicios de que el cráneo del "Hombre de Orce" pertenece a un asno', Relaño, *El País* 1984a.

Orce bone (see figure 8).[16] It is important to state that it will be very difficult to establish exactly how information flowed from Paris to *El País*, but it is crucial to highlight that things happened very fast, in a matter of a few hours from the discoverers' visit to the de Lumleys to the public announcement in the newspaper. Later, the *El País* scoop was followed by hundreds of news items about the Orce Man in the national and local press over the following twenty years, almost up until today.[17]

Alfredo Relaño was the *El País* journalist that first covered the Orce controversy story. Although in 1984 he dealt with all kinds of Andalusian general political news for *El País*, he was, and still is, mainly a sportswriter; specifically, a football journalist.[18] But the Orce story seemed too attractive to miss out on: an already very well-known science story turning into a controversy, with implications for local politics and with the possibility of lasting some time. It was a perfect 'conflict story' for the journalist and for the newspaper that put it on its front page.[19] Relaño was then working in the Andalusian office of *El País* and it seems that the Orce story reached him by chance. According to Relaño's subsequent statement, he did not receive the news from the newspaper but from the *Junta de Andalucía* directly.[20] Yet, in Relaño's articles, three main sources can be spotted: Javier Torres Vela, head of the *Junta de Andalucía*'s Department of Culture; Isidro Toro, an Andalusian archaeologist from the Venta Micena research team; and "a source of the most reliable kind", Marie-Antoinette de Lumley. So, although it is not clear if the *Junta* acted as a mediator, what is clear is that Relaño communicated with "Marie-Antoinette de Lumley last night in a phone conversation with this newspaper".[21]

As Relaño explains, the little crest first confused the discoverers but later "made them fear that [the fossil] may not be a human ancestor", adding, "it presents all the evidence of being an equine similar to the present day ass."[22] Toro stated that there was a 20 per cent probability that the Orce Man was in fact a young ass. It is noteworthy that Relaño used the word *asno* (ass), never *burro* (donkey), which in Spanish is the more derogatory term. Like subsequent

16 Relaño, *El País* 1984b; Redacción, *El País* 1984a; Redacción, *El País* 1984d.

17 For more on the media attention to the Orce controversy see Annex 2.

18 Wikipedia Contributors 'Alfredo Relaño'.

19 See Miller/Gregory 1998, 110–114.

20 Navarro 2010.

21 "Fuente de todo crédito", "Marie-Antoinette de Lumley en conversación telefónica con este periódico …", Relaño, *El País* 1984b.

22 "Les ha hecho temer que no sea un antepasado del hombre", "tiene todos los indicios de ser un équido similar al asno actual", Relaño, *El País* 1984b.

24 LA CULTURA EL PAIS, sábado 12 de mayo de 1984

Dudas sobre la validez de un hallazgo histórico

En junio de 1983, en la localidad granadina de Orce, paleontólogos andaluces y catalanes anunciaron lo que podría ser uno de los descubrimientos paleontológicos más importantes de los últimos años. La aparición de un trozo de cráneo fosilizado hizo pensar que se trataba de un hombre de una antigüedad entre 900.000 y 1.600.000 años, lo que le convertía en el resto humano más viejo encontrado en Eurasia. Casi un año después, los paleontólogos encuentran serias dudas sobre el origen humano de esos restos y han suspendido un simposio en el que se iba a estudiar el hallazgo. Las últimas investigaciones hacen sospechar que, en realidad, el Hombre de Orce podría ser un asno de dos a cuatro meses de edad.

Serias sospechas de que el 'hombre de Orce' no es un hombre, sino que podría ser un asno de dos a cuatro meses de edad

ALFREDO RELAÑO, **Sevilla**

El llamado *hombre de Orce*, hallado en la localidad granadina que le da nombre a este supuesto hallazgo histórico, podría ser un ásnido de dos a cuatro meses de edad, según indicios razonables descubiertos tras el estudio del endocráneo, una vez limpio éste de la caliza fosilizada que tenía adherida. Esta posibilidad ha movido al aplazamiento del Simposio Internacional convocado en Granada para finales de este mes, y que tenía el carácter de presentación oficial del hallazgo ante las máximas autoridades mundiales en la materia. El aplazamiento, cuya responsabilidad política asume la Junta de Andalucía, ha sido decidido a petición del propio equipo descubridor del hallazgo, cuyo director, Josep Gibert, se desplazó el miércoles a Sevilla, junto a dos miembros de su equipo, para solicitar el aplazamiento.

MARCEL·LÍ SANZ

El parietal del llamado 'hombre de Orce' con sus descubridores, los paleontólogos del Institut de Paleontologia de Sabadell. De izquierda a derecha, Josep Gibert, Jordi Agustí (con el parietal en la mano) y Salvador Moya.

El *hombre de Orce*, un pedazo de cráneo atribuido en principio a un humanoide, fue hallado en Venta Micena, en el término municipal de la localidad granadina de Orce, por el equipo del Instituto de Paleontología de Sabadell, dirigido por el profesor Josep Gibert. Tras su examen por Miguel Crusafont, una máxima eminencia en la paleoantropología española, ya fallecido, fue interpretado como el vestigio humano más antiguo encontrado en Eurasia, como un tránsito entre el *homo habilis*, hallado por el matrimonio Leakey en la gargante de Olduvai, en Kenya, y los numerosos vestigios de *homo erectus* encontrados en Europa y Asia. La condición de homínido del resto del fósil le fue atribuida, básicamente, y según numerosos expertos consultados por este periódico, por el dibujo de la fisura entre los dos huesos del cráneo que constituían el resto.

Conclusiones controvertidas

Las conclusiones del profesor Miguel Crusafont fueron discutidas por otros paleoantropólogos, entre ellos, Emiliano Aguirre, máxima figura de la *escuela de Madrid*, pero sus dudas fueron atribuídas a recelos científicos entre dicha escuela y la *escuela catalana*, liderada por Miguel Crusafont. Un viaje a Orce del matrimonio formado por Marie Antoinette y Henry de Lumley, del Museo de Historia del Hombre de París, y tras el que no se recataron en afirmar que el de Venta Micena era el más importante yacimiento de vestigios humanos de Eurasia, terminó de dar respaldo oficioso al hallazgo. Todo ello movió a la Junta de Andalucía a convocar un Simposio Internacional, que debería haberse celebrado en Granada los próximos días 28, 29 y 30 de este mes, y al que estaban invitadas las máximas autoridades mundiales en Preshistoria, Antropología y Paleontología, para presentar de forma oficial en el mundo científico el hallazago, cuya importancia se consideraba suficiente en todos los círculos relacionados con la paleontología humana como para no ser presentado en un congreso ordinario, sino merecedor de un acto así.

El problema apareció cuando, tras la paciente tarea de limpiado de la cara interna del cráneo, oculta por una capa de caliza fosilizada, se ha descubierto en la misma una pequeña cresta, que en principio desorientó a los responsables del hallazgo, y que después, y según fuentes de todo crédito a las que ha podido tener acceso este periódico, les ha hecho temer que pueda no tratarse de un antepasado del hombre y que, aunque se intenta relacionarlo con alguna rama de homínido desaparecida, tiene todos los indicios de ser un équido similar al asno actual.

Una nota lacónica de suspensión

El problema que una confirmación de esta posibilidad posterior a un Simposio Internacional podría plantear, movió a tres miembros del equipo descubridor, el director del mismo, Josep Gibert, director del Instituto de Paleontología de Sabadell y máximo responsable del hallazgo, Isidro Toro, investigador del Museo Arqueológico de Granada, y Pascual Rivas, catedrático de Paleontología de la Universidad de Granada, a desplazarse a Sevilla para solicitar de los responsables de la Consejería de Cultura de la Junta el aplazamiento del simposio, que ya estaba convocado y con las invitaciones cursadas.

La Consejería de Cultura de la Junta accedió a la petición y pasó una lacónica nota a los medios de comunicación, en la que se informa de un aplazamiento del simposio, "a petición del equipo investigador". El consejero de Cultura, Javier Torres Vela, que hace pocas semanas llegó al cargo, se limitó a asumir la responsabilidad política del aplazamiento y a reafirmar que éste se hacía a instancias del equipo investigador, ante la existencia de dudas sobre el hallazgo. "No se duda de su antigüedad, sino de que se trate efectivamente de un homínido". Enrique Vallespí, profesor de Prehistoria de la Universidad de Sevilla, y a quien se había encomendado la coordinación científica del simposio, criticó ayer, en declaraciones a este periódico, tal decisión. "Aunque existan esas dudas, debería mantenerse el simposio, porque era una ocasión única de realizar un estudio profundo, una puesta al día, de los primeros pasos del hombre en Europa".

Un viaje científico a París

Según pudo saber este periódico, en el aplazamiento tuvo una importancia decisiva el paso atrás dado por el matrimonio Lumley, hasta ahora tan entusiasmado con el hallazgo que el Museo de Historia del Hombre de París ha montado una exposición sobre los primeros pasos del hombre en Europa, en la que era figura estrella el hallazgo de Orce. La exposición, con carácter itinerante, iba a hacer el recorrido París-Niza-Perpiñán-Madrid. Según nuestras informaciones, fue la investigadora francesa Marie Antoinette de Lumley quien aconsejó a Josep Gibert la suspensión del simposio, e incluso llegó a decirle que si no se realizaba tal suspensión ella y su marido pretextarían una enfermedad para no acudir.

Marie Antoinette de Lumley recibió del equipo de Gibert un molde del endocráneo del *hombre de Orce*, y fue ella quien, tras un estudio de su equipo, recomendó compararlo con el de asnos de dos a cuatro meses de edad. De hecho, el equipo de Gibert ha recopilado en las últimas fechas cráneos de asnos jóvenes a fin de estudiar las suturas interiores de los huesos.

El profesor Gibert se encontraba ayer en París, estudiando con el matrimonio Lumley, fuertemente comprometido también con el hallazgo, las posibles salidas a la cuestión. Isidro Toro, único miembro del equipo con el que este periódico pudo entrar en contacto ayer, admitió que existía un 20% de posibilidades de que no se tratara de un hombre, pero expresó su esperanza en la falta de estudios sobre endocráneos de esa antigüedad y en el hecho de que se pudiera descartar firmemente que se trate de cualquier tipo de carnívoro y de un ciervo.

No descartó la posibilidad de que se trate de un ásnido, pero insistió en que todavía cabe esperar que el resto hallado proceda de algún primate.

Pese a la probable frustración del hallazgo, que ha movido a que la Junta de Andalucía envíe hoy 500 cartas en suspensión de otras tantas invitaciones al anunciado simposio, diversos medios consultados defienden la importancia del yacimiento de Venta Micena, en el que, aun en el caso de que se descarte definitivamente el carácter excepcional atribuido inicialmente al fósil, se han encontrado vestigios de industria humana que se remonta a la frontera entre el Pleistoceno inferior y el Pleistoceno medio; tan es así, que el equipo del Instituto Paleontológico de Sabadell va a trabajar con intensidad en los próximos meses de verano.

El hecho de que el equipo descubridor haya pedido el aplazamiento del simposio durante seis meses y no la suspensión definitiva permite suponer que mantiene esperanzas de encontrar algún resto de homínido en este plazo de tiempo. Por otra parte, Josep Gibert mantiene igualmente la convocatoria de una conferencia que piensa pronunciar en Madrid el próximo día 17, y que tenía el carácter de antesala del simposio aplazado.

FIGURE 8 *El País* published the scoop on the possible equine classification of the Orce Man
FROM RELAÑO, *EL PAÍS* 1984B

news pieces, the article highlighted the significance of the Venta Micena site, even if it was finally demonstrated that the bone was not from a hominid. Javier Torres Vela provided Relaño with details about the postponement of the international conference, with regard to which the articles mainly highlighted the five hundred cancellation letters sent by the *Junta* to national and international scientists. According to Relaño, as well as the discoverers team, Marie-Antoinette herself also recommended the conference's cancellation to the *Junta*.[23]

During the following days, *El País* published different articles that accused politicians from the *Junta* of using the Orce Man discovery for their own political advantage, a critique that was hardly made a year before when the discovery was announced and the controversy had not yet broken out. According to one journalist, politicians "did not miss the opportunity to translate science into politics" and the conference was yet another opportunity that served for taking political advantage.[24] Relaño himself hinted at the possible political erosion that this cancellation could cause in the *Junta*.[25] It appeared to journalists that responsibility for the political damage caused by the Orce mistake and the cancellation of the conference must be assumed by these politicians.[26] To defend themselves, the *Junta* politicians assumed the responsibility but stated that they were always, both with the presentation of the discovery and with the cancellation of the conference, following the scientists' criteria. Thus, if someone were at fault here, it was not them but the discoverers.[27] At the same time, Gibert defended himself by stating that they never asked for a big conference, just a small meeting to show the discovery to a limited number of international palaeoanthropologists, and that they did not request the cancellation of the conference, merely its postponement.[28] In turn, the *Diputació de Barcelona*, which oversaw the *Institut de Paleontologia*, expressed its support for the discoverers and their official position that the bone belonged to a hominid. It was also stated that the proper study of the bone would not be ready until the following December, so until then no final decision about it would be reached.[29]

23 Relaño, *El País* 1984b.

24 "No desaprovecharon la ocasión para traducir la ciencia a la política", Redacción, *El País* 1984b.

25 Relaño, *El País* 1984c.

26 See for instance: Relaño, *El País* 1984c and Redacción, *El País* 1984a.

27 Relaño, *El País* 1984c.

28 Redacción, *El País* 1984c; EFE, *ABC* 1984a; J. J. N. A., *El País* 1984.

29 Interview with Roger Marcet 2013; Interview with Campillo 2011; EFE, *ABC* 1984b; Guerrero, *La Vanguardia* 1984; Redacción, *La Vanguardia* 1984c.

According to Gibert, when he came home from Paris lots of phone calls from journalists and friends were awaiting him.[30] During these first days or even first years after the *El País* front page, the main voice of the discoverers was his. Moyà-Solà and Agustí remained in the background almost completely while Gibert took the leading role. According to them, the two junior scientists did not want to publicly contradict their superior, as Gibert was not only their senior but also acting director of the *Institut de Paleontologia de Sabadell*.[31] Or maybe they were just waiting and watching how the story developed. In any case, as Gibert already took this leading position during the early non-controversial popularisation of the finding, he remained there as the most convinced of the three that the Orce bone was not an equine remain.

Gibert's discourse followed three main lines of argument during the first days of the controversy. Firstly, he stated that the most likely possibility was still the hominid option. To support this, he spread around the media some very technical explanations on the internal features of the bone and its comparison with the Parisian specimens (see also figure 7):

> [The crest] in our cranium is single, in horses it's double. […] The differences in the crest are the following: it's double in the IPH [*L'Institut de Paléontologie Humaine*] [horse] specimen, and single in the young [onager] specimen from the LAC (Comparative Anatomy Laboratory), but in both cases, it has a shape that enlarges at the base, never the flat shape of the Orce Man.[32]

These same comparisons were later used in a scientific conference that Gibert attended soon after the controversy started (see below).[33] In addition, Gibert protected his own reputation as a scientist adding that if they were wrong, then all of the internationally recognised experts that had seen the bone before the cleaning process, including the de Lumleys, were also wrong.[34] The second issue that Gibert raised was that the significance of Venta Micena was independent of the features of the Orce Man bone.[35] Here Gibert protected

30 Gibert 2004, 49–50.

31 Interview with Agustí 2012 and Interview with Moyà-Solà 2012.

32 "[La cresta] en nuestro cráneo es sencilla, los caballos la tienen doble. Las diferencias de la cresta son las siguientes: Es doble en el ejemplar del IPH es sencilla en un ejemplar joven del LAC (Laboratorio de Anatomía Comparada), pero, en ambos casos, presenta una forma que se agranda hacia la base, nunca la forma laminar del hombre de Orce", Arroyo, *El País* 1984. For more on the features of the Orce fragment see Annex 1.

33 Gibert 1984.

34 Arroyo, *El País* 1984.

35 *Ibid.*

his own research project from the possible damage that the public controversy may cause it. Finally, he also pointed to "extra-scientific issues", "exaggerated rumours", or even "dubious interests" as the reasons why the crest problem had reached the media.[36] With these statements, Gibert, who apparently had always wanted to remain on "scientific terrain", clearly tried to distance himself from those that had started and followed the controversy in the media, mainly the de Lumleys and the *Junta* politicians.[37] In the same vein, Salvador Moyà-Solà (in one of his few media appearances during the early controversy in the local Sabadell newspaper) remained silent about through whom, how, and why the controversial information had reached the *El País* front page, stating only that "it seems that somebody is interested in discrediting our work".[38] Moyà-Solà also followed Gibert's defensive line of argument by saying that the discovery was backed by several international experts, claiming also that they wanted to remain within "scientific boundaries".[39] This interview shows us how during the first days of the controversy, at least Moyà-Solà was presenting a similar discourse to that of Gibert.

To some extent, the controversy also had an international echo. The British *The Daily Telegraph* and *The Times*, the Canadian *Ottawa Citizen*, and the American *The Gadsden Times* covered the *El País* scoop and highlighted the cancellation of the conference. These newspapers also noted that the mix-up had specifically been made with an ass or donkey using headlines like: 'Ass is taken for a man' or 'Scientists make ass of an old skull'.[40] Moreover, international newspapers highlighted the need for further investigation to be carried out to clarify the fragment's species and even the "dubious interests" that promoted the donkey idea. Most of these newspapers followed the EFE agency's news release, although the British also had their Madrid correspondents.

To sum up, during these first controversial days, the dispute was not one between those who defended one scientific position and those who defended another, but was mainly about who assumed the responsibility (both scientific and political) of the supposed 'Orce mistake' and the conference cancellation. First, politicians protected themselves by placing the responsibility for their actions on scientists. Second, Agustí and Moyà-Solà successfully protected

---

36 "cuestiones extracientíficas", "rumores magnificados", "intereses oscuros", EFE, *ABC* 1984a.

37 "terreno científico", J. J. N. A., *El País* 1984.

38 "Parece que alguien está interesado en desprestigiar nuestro trabajo", Colomer, *Diari de Sabadell* 1984.

39 "limites científicos", *Ibid.*

40 Debelius, *The Times* 1984; Our Madrid Correspondent, *Daily Telegraph* 1984; Editorial department, *Ottawa Citizen* 1984; Editorial department, *Gadsden Times* 1984.

their reputation by 'hiding' from the public arena and giving the lead voice to Gibert. And the latter tried to construct a shield to protect himself, his research, and his scientific reputation from the harm that the public display of the doubts around the Orce bone could have caused, or would cause. Together with Gibert, everyone seemed to try to protect the Venta Micena site and its significance. Meanwhile, all of this was accompanied by technical information about the factors that could lead scientists to decide whether the bone was a hominid or not.

It is also worth emphasising that without the great popularisation effort made by Gibert, Agustí, Moyà-Solà, and the politicians during the first year after the discovery, the de Lumleys' opinion would not have received so much attention. A path had been paved for the public controversy that would not have existed without previous public interest in the case.[41] When Relaño or someone at the Andalusian *El País* office received the news, he or they already knew about the Orce Man and its significance. Moreover, they knew that the news that came from Paris were a very interesting scoop, worth to be at the front page.

Finally, it is also crucial to state that during that early phase, the main historical actors themselves (the discoverers, the politicians from the *Junta*, and the de Lumleys) were the ones who decided what information became public and what was kept private. It is clear that Gibert's first option was not to go public regarding the doubtful crest but to continue working on the Orce fragment privately, making only uncontroversial statements publicly. He also refused to hold a press conference to admit their mistake. The move of making the controversy public was not his but the de Lumleys' and/or the *Junta* politicians'. The French couple, instead, did not want a private or a public debate, neither wanted the research around the Orce man to keep going without any statement, but they did want to establish their opinion clearly, with no doubts, and this had to be done publicly.

This will of the de Lumley couple and/or of the *Junta* politicians is what made possible the 'Orce Man public metamorphoses'. From a 'public scientific discovery' to a 'public scientific controversy'.

∴

On 12 May 1984, the controversy broke out in *El País*. The next day, the Catalan newspaper *El Periódico* published an opinion piece entitled 'We descend from

41 See a similar case in Goulden 2013.

the donkey', which mainly used the Orce story to criticise present-day human attitudes and activities like wars or football hooliganism.[42] That was one of the first articles that used *burro* (donkey) in relation to the Orce Man instead of *asno* (ass), which had been used up until then. *Burro* was then first used as a metaphor to criticise certain attitudes rather than to criticise the discoverers or their research. On 15 May, Gibert, carrying the Orce bone, attended the opening of an exhibition organised by the Parisian *Musée de l'Homme* in Perpignan that was also attended by Henry de Lumley.[43] De Lumley stated to the Spanish press that the possibility that the Orce bone was a hominid "cannot be dismissed". He also described the *Institut de Paleontologia de Sabadell*'s scientific team as "one of the most important in the world", and, finally, he praised the Venta Micena site and the Orce region.[44] After dropping the bomb, the de Lumleys did not want to further extend the public conflict. His authority was secured once again. The controversy was not worth continuing as ultimately it could affect them too.

That same day, *El País* published an opinion piece entitled 'Hombre o Borrico' ('Man or Donkey').[45] Again, this article used 'donkey' to maximise a critique that was not aimed at the scientists but at the politicians that used a scientific discovery to their own advantage. The next day (16 May), *La Vanguardia* published another opinion piece entitled 'El Burro de Orce' ('The Orce Donkey'). This time, the journalist commented that it seemed that the donkey, an animal always related with ignorance, was chosen on purpose to insult the discoverers and their scientific effort.[46] For him, if the bone had belonged to a monkey, or maybe a horse, or a bear, the controversy would not have been so major. Yet so far, this notion of the Orce donkey had not been explicitly used against the discoverers. That same day, *El Periódico* published a comic strip in which the cartoonist presented two scared scientists in a hole with the Orce bone surrounded by several apparently hungry donkeys. The critique was again directed at politicians and journalists for their excessive and exaggerated interest in the finding.[47]

---

42 'Descendemos del Burro', Pernau, *El Periódico* 1984.

43 J. J. Navarro Arisa, *El País*, 1984 and Redacción, *El Periódico* 1984b. More on the 'travelling bone' in these early years in Annex 3.

44 "no puede ser descartada", "uno de los más importantes del mundo", J. J. Navarro Arisa, *El País* 1984.

45 Redacción, *El País* 1984b.

46 Gomis, *La Vanguardia* 1984.

47 Morales, *El Periódico* 1984.

Also on 16 May, Gibert gave a talk in the *Museu d'Arqueologia de Catalunya* to an audience "hungry for news" about the bone.[48] The next day, he went to Madrid to give another lecture that had been scheduled before the controversy started.[49] According to Gibert, when he arrived at the venue, the *Museo Arqueológico Nacional*, journalists "assaulted" him, but he was "saved" by Eduard Ripoll, the museum director.[50] There Gibert gave a talk in front of a large audience with several scientists, including Emiliano Aguirre, one of the most well-known Spanish palaeoanthropologists who had started the Atapuerca research project in 1978.[51] According to Gibert, at some point during this period he also went to Orce to give a talk to the "very interested audience" of the town's "many citizens", including the mayor of Orce, José María González Galera.[52] As Gibert stated, "with the controversy, the lecture requests and the need to provide explanations intensified".[53] In June, Gibert went to a scientific conference in memory of the 50th anniversary of the death of Luis Siret (1860–1934), a Belgian mining engineer and archaeologist who had worked in south-eastern Spain. Since this was a large conference, it can be assumed that Gibert's talk was scheduled before the controversy started. The title of his talk was 'The Venta Micena site (Orce, Granada): Its importance and anthropogenic action, and the palaeoanthropological characteristics of the *Homo* sp. cranial fragment'.[54] This title also supports the assumption that the talk was scheduled before the controversy since it was very general and nothing was mentioned about the controversial crest. In spite of that, Gibert presented to a "crowded room" the study of the anatomical characteristics of the bone's inner part and its comparison with animals and hominids.[55] He used data compiled in Paris that he had already made public in several media appearances. Precisely at that point, it seemed that popular and scientific media, popular and scientific audiences, and, thus, popular and scientific discourses were completely mixed.

Also in June, the satirical magazine *El Papus* published a front cover showing a cartoon of a prehistoric man with a donkey's head and the title 'The Orce Man Mystery' (figure 9).[56] The cartoon speech bubble read "I'm Spanish, big

48 "ansioso de noticias", Gibert 2004, 55.
49 *Ibid.*, 55. AJG-ICPMC: Museu Nacional de Arqueologia de Madrid leaflet, 1984.
50 "asaltaron", "salvado", *Ibid.*, 55.
51 *Ibid.*, 55.
52 "muy interesado", "numerosos ciudadanos", *Ibid.*, 53.
53 "Con la polémica se intensificaron las peticiones de conferencias y la necesidad de dar explicaciones", *Ibid.*, 54.
54 'El yacimiento de Venta Micena (Orce, Granada). Su importancia, acción antropogénica y características paleoantropológicas del fragmento de cráneo de *Homo* sp', Gibert 1984.
55 "sala repleta", Gibert 2004, 54.
56 Redacción El Papus 1984.

deal!" in a very colloquial, rural, even uneducated Andalusian accent.[57] This prehistoric Spanish donkey also had an "I love Fraga" tattoo. Manuel Fraga Iribarne (1922–2012) was a right-wing Spanish politician who had been a minister during Franco's dictatorship and who at that time was the leader of the political party *Alianza Popular* (AP), which represented right-wing ideas after the dictatorship.[58] This front cover was also a satirical critique of the Spanish political, social, and maybe scientific situation, and of Fraga and his party's ideas, rather than of the scientists or their mistake. For the cartoonist, the Orce Man story was ideal for linking Spanish backwardness, and especially the situation in rural Andalusia, to conservative former-Francoist politics and politicians. In spite of that, this cartoon is sometimes portrayed as an assault on Gibert's credibility.[59]

To sum up, Gibert presented his opinions (outlined in the previous section) to a wide range of different 'publics' and through a great mixture of communication channels and discourses. From the scientific audiences of the Siret conference or the *Museo Arqueológico* to the lay audience of Orce, Gibert was expanding and strengthening his position in the controversy. The public impact of the *El País* front page had to be tackled, and Gibert had done just that. Contrary to Gibert's later discourse, the analysis of how the notion of the 'Orce Donkey' emerged in the Spanish media shows how it appeared little by little and usually not to criticise Gibert or the discoverers but to criticise politicians and Spanish society. The 'donkey' was too appealing a notion for journalists not to use it against the political class. As the de Lumleys' statements in Perpignan also show, after the first announcement in *El País* neither Gibert nor his work received many more 'attacks', at least for the meantime.

2.2 A Country's 'Obsession': 'Is the Orce Man our Ancestor?'

As expected, everybody once again jumped on the Orce bandwagon after the publication of Relaño's article. Unlike the first time, it was now a controversy bandwagon. Journalists gave voice to several different 'actors' that carried weight in this controversy. For instance, Relaño himself went to Orce to ask the town's inhabitants about their thoughts.[60] The *Junta* politicians replied

57 "Zoy Ezpañó, cazi ná!", *Ibid.*

58 Wikipedia Contributors 'Manuel Fraga Iribarne'.

59 See, for instance, Lluís Gibert's statements in: Navarro 2010. Also: J. A., *Diari de Sabadell* 1986c.

60 Relaño, *El País* 1984d.

FIGURE 9 *El Papus* front cover, June 1984, from Redacción *El Papus* 1984
FROM REDACCIÓN *EL PAPUS* 1984

to criticisms in newspapers, scientists such as Toro or de Lumley were asked about the controversy in their public appearances, and Gibert took the lead on behalf of the discoverers. Yet, apparently, experts could not reach an agreement, but the country wanted to know: is it a donkey or a hominid?

In the summer of 1984, the *Institut* team went to Orce to excavate. In spite of the controversy, the project doubled its numbers. More excavators and more days were allowed due to the renewal of the agreement between the *Junta* and the *Diputació*. Still, Gibert was quoted in the press complaining about the excavation project's financial situation.[61] Newspapers presented Gibert as the leader of the Sabadell team. His name made the headlines, while Agustí's and Moyà-Solà's names barely appeared.[62] Gibert's early popularisation effort and his 'defence' after the controversy had turned him into the main character of the Orce story.[63] Before the excavation started, Gibert stated that he was hoping that that year's research would confirm the human presence in Venta Micena, a notion which appeared clear before the controversy, when Orce was a success story and had not needed any further confirmation. During the excavation, Gibert stated in ABC that finding a hominid was "really an obsession" for him.[64] In the same article, the journalist stated that both Gibert and Isidro Toro were convinced that sooner or later they would confirm human activity at the site.[65] According to the press, during this period the French scientist François Sémah, a disciple of Henry de Lumley and recommended by him, went to Orce to carry out palaeomagnetic studies on the Venta Micena site.[66]

After the summer, in October 1984, the famous Galician punk band Siniestro Total published a song called 'Who we are? Where do we come from? Where are we going?', in which, among other questions, the songwriter Javier Hernández wondered: "Is the Orce Man our ancestor?"[67] The song later became one of Siniestro Total's most popular songs, even providing the name for their 2002 compilation album.[68] Several cartoons were also published in the first months after the controversy. As discussed earlier, most of them did not criticise the researchers or their research but mainly used the 'Orce donkey' in their critiques of other, mostly political, issues. In *El Jueves*, a well-known Spanish satirical magazine, the Orce Man did not make it onto the front cover but did have a full-page article and another short article. They

61 Vilardebó, *ABC* 1984.

62 Avila Granados, *La Vanguardia* 1984 and Otr., *ABC* 1984.

63 For example: Otr., *ABC* 1984.

64 "ciertamente una obsesión", Vilardebó, *ABC* 1984.

65 *Ibid.*

66 Avila Granados, *La Vanguardia* 1984. For more on Sémah: Anonymous 'François Sémah'.

67 '¿Quiénes somos? ¿De dónde venimos? ¿A dónde vamos?', "¿Es nuestro antepasado el Hombre de Orce?", Hernández 1984.

68 Siniestro Total 2002, '¿Quiénes somos? ¿De dónde venimos? ¿A dónde vamos?'.

mostly used the joke that humans now came from donkeys and no longer from monkeys. 'Great joy among monkeys' was the title of the full-page article, which commented on the monkeys' statement: "It's great news for us to know, once and for all, that we have nothing to do with humans."[69] The short article ironically linked the Orce Man to the increasing nuclear tensions between the USSR and the United States by stating that "humans, and especially Chernenko and Reagan, may descend from donkeys".[70] In *El Periódico*, the well-known Catalan cartoonist El Perich also joked about the global political situation. He drew two aliens arriving to Orce in the future: "It appears that it was not a man ...it was a donkey!" says one alien with a bone in its hand, adding, "He died in a nuclear war!"[71] These examples, like the *El Papus* cover and the Siniestro Total song, show the degree to which the controversy had seeped into the Spanish public sphere and popular imagination. Several actors appropriated the Orce Man scientific controversy in their commentaries on the world as something that was a given, something known to everyone and that did not require any further explanation. As mentioned earlier, some of these examples were used to demonstrate the harsh accusations that Gibert and his research received in these early years of the controversy. Yet, in most of these instances the Orce Man's popularity and the appealing notion of our being of 'donkey descent' were used in the authors' social and political critiques.

At the end of 1984 and the beginning of 1985, Gibert publicly announced three lines of research to the press that aimed to demonstrate that the Orce bone was a hominid: firstly, comparing the Orce fragment with remains of different ancient animals for which he had started to work with PhD students; secondly, comparing the fragment with present-day human features, here, Domènec Campillo, a neurosurgeon that saw the bone in the beginning, was helping Gibert with a comparative study; and thirdly, the possibility of doing a biochemical analysis of the bone, which would provide 'definitive' results and for which he had already contacted scientists from the *Universidad de Granada*.[72]

The way that the press dealt with the Orce story and covered the excavation, the way that the 'Orce Donkey' notion was introduced, the fact that money

---

69 "Gran júbilo entre los monos", "Es una gran noticia para nosotros saber de una vez por todas que no tenemos nada que ver con el hombre", Redacción El Jueves 1984.

70 "el hombre, y sobre todo Chernenko y Reagan, descienden tal vez del asno", Romero 1984.

71 "Parece que no fue un hombre, ¡fue un burro!", "¡Murió en una guerra nuclear!", El Perich, *El Periódico* 1984.

72 EFE, *La Vanguardia* 1985; Interview with Campillo 2011; Campillo 2002, 33.

from institutions was still available, and the visit of the French specialist from the de Lumleys' laboratory show how, at that point, Gibert, the Orce Man, and the research in Venta Micena were still not marginalised or ignored in the public or in the scientific sphere. In a way, Gibert merely tried to respond to the many questions from the press and to continue with a popularisation effort similar to the one present before the controversy. Gibert has often been publicly accused of being 'obsessed' with the Orce bone.[73] After his reaction to the de Lumleys' verdict, the development of his studies on the bone and his statements in the press suggest that perhaps it was at that time that Gibert began to develop this 'obsession'. Yet, the Orce Man was a hot topic during this period in the Spanish public sphere and despite there not being much criticism raised specifically against the Orce bone after the *El País* front page, the country wanted to know more, the whole country was 'obsessed', and Gibert was there to answer (even if it was his very own personal reply).Thus, the supposed 'obsession' did not develop in the first years after the controversy broke out, but was a subsequent 'construct' involving several actors.

∴

In order to look for support, in 1985, Josep Gibert contacted the immunologist Enrique García-Olivares from the *Universidad de Granada* and suggested he perform a technique for identifying fossil proteins in palaeontological remains on the famous cranial fragment. Gibert brought the work of the American immunologist Jerold Lowenstein to García-Olivares. Since the early 1980s, Lowenstein had been performing this technique on different fossils, such as mammoths, and publishing them in high-profile journals like *Science* or *Nature*.[74] On his own website, Lowenstein defines himself as "a pioneer in the study of fossil molecules" as he "opened a new field of research, fictionalized in the book and movie *Jurassic Park*".[75] Lowenstein's research was part of a wider movement in palaeontology and anthropology that took place between the 1960s and the 1980s. As historian Marianne Sommer argues, this research was turning molecules, and especially genes, into "the most fundamental, clean, and direct records of historical [biological] information."[76]

73 Antón, *El País* 2000; Planas, *La Vanguardia* 2007; Interview with Agustí, 2012 and Interview with Moyà-Solà, 2012.

74 Interview with Enrique García-Olivares 2014; Gibert 2004, 64; Prager et al. 1980; Lowenstein 1981a and Lowenstein/Sarich/Richardson 1981b.

75 Lowenstein 2013.

76 Sommer 2008, 473.

This caused a reaction from traditional palaeontologists that defended the validity of remains themselves within these sciences. Despite problems of contamination, the popularity of these techniques, especially ancient DNA research, has grown since the 1980s and they have been applied with success (and controversy) over the years to remains as old as those of 65 million-year-old dinosaurs.[77]

Yet, in these cases, the technique was applied mainly to explain the evolutionary origin of specific remains. Instead, following Lowenstein's work, García-Olivares performed an analysis to search for human albumin in VM-0 in order to establish its classification. This was the first time that such a technique was applied in Spain and in Europe. Later, Lowenstein himself attended the Orce conference, took samples of the Orce bones, and performed the analysis in the United States.[78] In both cases, the result of the analysis was that the controversial Orce Man did in fact present human albumin. In short, despite the fact that the search for human albumin in hominid fossils has not been performed much since, apparently this kind of technique has been used by several research teams and published in the most important scientific journals. García-Olivares himself stated that he never received criticism from the immunology community, only from palaeontologists. For him, the fact that proteins had been recovered from 65 million-year-old dinosaurs meant that Lowenstein's and his own results could definitely support the claim that the Orce Man was a hominid.[79]

At the end of January 1985, the press announced the discovery of a *Homo* sp. phalanx in the Cueva Victoria cave in Murcia (see map 1). This was the cave that already in 1981, before the Orce bone was found, was labelled as the oldest human site on the Iberian Peninsula in an article signed by Agustí, Moyà-Solà, Joan Pons-Moyà, and Eudald Carbonell, among others.[80] Pons-Moyà, an amateur palaeontologist and collaborator of the *Institut de Paleontologia de Sabadell*, found the phalanx and first published it in a scientific paper, again in *Endins*, the journal of the *Federació Balear d'Espeleologia*. Using biostratigraphic methods, the phalanx was dated at between 1 and 1.5 million years old. Therefore, it "may be one of the oldest examples of our genus in Spain."

---

77 Hedges/Schweitzer 1999 or Hedges/Schweitzer 1995. For criticisms see for instance: Fiedel 1996.

78 Interview with Enrique García-Olivares 2014; Borja et al. 1992; Borja 1995; Borja et al. 1997; Lowenstein/García-Olivares/Borja 1999.

79 Interview with Enrique García-Olivares 2014. See for instance: Schweitzer et al. 2009.

80 Moix, *La Vanguardia* 1985; Barata, *El Periódico* 1985; Contreras, *El País* 1985. Carbonell et al. 1981.

According to Gibert, Pons promised to allow his team to study the phalanx later.[81]

Although to a lesser degree, the public presentation of the Cueva Victoria phalanx did repeat some trends from the Orce Man presentation. Someone from the *Institut de Paleontologia* sent a letter to Murcia's Department of Culture to announce the discovery of remains that were "100,000 years older than the Orce Man". This somehow 'jumped' into the media. Jordi Agustí stated to the press that this phalanx could be "the oldest known human remain in Europe". Gibert communicated with *El Periódico*, which devoted two whole pages to the discovery. Again, with this new discovery, newspapers highlighted the precariousness of the palaeontological research carried out by Gibert and his collaborators.[82] The Cueva Victoria phalanx was somehow 'appropriated' by Gibert to support his ideas of the early human occupation of southern Spain and, therefore, to reinforce the Orce research, damaged after the *El País* front cover (figure 8). As he stated in *La Vanguardia*, "[this phalanx] confirms our lines of research and increases the significance of the Orce discoveries".[83] Again, a collaboration agreement between the government of Murcia and the *Diputació de Barcelona* was about to be signed since the research project in Cueva Victoria was going to begin in March 1985.[84] Nevertheless, this new public presence also raised problems for Gibert.

In March 1985, three palaeontology professors (all disciples of Miquel Crusafont just like the Orce Man discoverers) sent an open letter to the Spanish media criticising Gibert and his methods. These professors were Jaume Truyols, from the *Universidad de Oviedo* and who has already appeared in our story; Miquel de Renzi, from the *Universitat de València*, and Jaume de Porta, from the *Universitat de Barcelona*. In their letter, the professors accused Gibert of using "unscientific methods" like the "reckless dissemination" of the findings' conclusions, described as "conjectures", which could lead the public to form "an absolutely erroneous opinion" of how palaeontology and science work.

For the professors, Gibert's problem was a procedural one: science acts with caution in its conclusions and its comparison of hypotheses, while Gibert did not. They concluded that proper use of the scientific method was "the

81 "puede ser una de las referencias más antiguas de nuestro género en España", Pons Moyà 1985. Gibert 2004, 57.

82 "el resto humano más antiguo que se conoce en toda Europa", Contreras, *El País* 1985. Barata, *El Periódico* 1985 and J. Pérez Piñar, *El Periódico* 1985.

83 "confirma nuestra línea de investigación y amplía el significado de los descubrimientos en Orce", Moix, *La Vanguardia* 1985.

84 J. Pérez Piñar, *El Periódico* 1985.

difference between the palaeontologist and the mere amateur fossil collector".[85] Moreover, they stated that the recent discovery of the phalanx had been attributed to a hominid "with the same haste" as in the Orce Man case.[86] De Renzi also prepared a scientific report in which he took apart Gibert's points regarding the Orce bone. He argued that with the present data it was impossible to prove definitively whether the fragment was a hominid or not. Of course, this scientific report was presented to the media.[87]

Some days later, Gibert replied in the press stating that he consistently followed the scientific method and that the dissemination of his discoveries was "always supported by previous publications and contrasted by experts."[88] Gibert later claimed that this letter caused "terrible" damage to his credibility and was a "great setback" for the Orce research project.[89] He also linked the letter to later political decisions that affected his research: "No politician opposes three university professors."[90]

At least the three professors' letter apparently had an effect on the decisions taken by some politicians. A month after it was publish, in April 1985, the secretary of Murcia's Department of Culture sent a letter to Gibert in which he reported on the previous month's Department meeting. In this meeting it was agreed to "urge" Gibert to abstain from embarking on "uncontrolled and excessive" dissemination of the Cueva Victoria findings as this could cause damage to the institution's prestige.[91] This same document also stated that the agreement between the Autonomous Community of Murcia and the *Diputació de Barcelona* "had to be studied" by the Department of Culture before being definitively approved.[92] Something had changed. Two years ago, Catalan and Andalusian politicians organised a press conference to present the Orce remains and their agreement. Now, the politicians of Murcia were being very cautious about both public dissemination and the agreement itself. The political damage that the

85 "metodos poco científicos", "imprudente difusión", "conjetura", "un juicio absolutamente erróneo", Redacción, *El País* 1985 and "Aquesta és la diferencia entre el paleontòleg i el mer col·leccionista de fòssils afeccionat", De Renzi/Porta/Truyols, *Avui* 1985.

86 "misma precipitación", Redacción, *El País* 1985.

87 Redacción, *El Periódico* 1985b.

88 "siempre han sido avaladas por publicaciones previas y contrastadas por especialistas." Redacción, *El Periódico* 1985a.

89 "terrible", "gran atraso", Gibert 2004, 56, 57.

90 "Ningún político se opone a tres catedráticos." *Ibid.*, 56.

91 "instar", "incontrolada y excesiva", AJG-ICPMC: Letter from the secretary of Murcia's Department of Culture to Gibert, 10 April 1985. Some time later, in 1987, Murcia's Department of Culture sent another letter to Gibert with similar complaints: AJG-ICPMC: Letter from Miguel Ángel Pérez-Espejo Martínez to Gibert, 27 October 1987.

92 "someter a estudio", *Ibid.*

controversy around the Orce fragment could have caused Andalusian politicians seemed stronger than the political benefits of a new public presentation.

In the summer of 1985, Gibert's team again went to excavate in Venta Micena. The excavation was reported on in Spanish newspapers, which also quoted Gibert maintaining that the Orce Man was a hominid.[93] This would be the last excavation with the three discoverers together. Before it began, Isidro Toro and the *Universidad de Granada* group, led by Pascual Rivas, decided to end their collaboration with the Catalan scientists. In the press, reasons were not specified, although "a split of the scientific team" was announced.[94]

On 11 October 1985, another blow hit Gibert and his team. This time in *La Vanguardia*, one of the most widely read newspapers in Spain and especially in Catalonia. The newspaper reported that seven of Gibert's team's archaeologists, among them Bienvenido Martínez-Navarro and Lluís Gibert, Josep Gibert's son, had been evicted from Cueva Victoria by the *Guardia Civil* (the Spanish equivalent of the French Gendarmerie, a federal military-status police force) due to their lack of excavation permits. According to the news piece, Gibert stated that his team was working there with a "verbal permit" due to the delay in the signature of the institutional agreement. For him, scientific research "should not be delayed for purely administrative reasons."[95]

That same month, politicians from the *Diputació de Barcelona* removed Gibert from his role as director of the *Institut de Paleontologia de Sabadell* and appointed Agustí instead.[96] According to Roger Marcet (then technical head of the Department of Culture of the *Diputació* and present at the meeting in which this decision was made), the *Diputació* politicians, although they supported Gibert and his interpretation of the Orce fragment, "did not want scandals" and thought that Agustí was a better option for avoiding them.[97] Still, Gibert could remain as a researcher in the *Institut*. Gibert attributed this decision to the professors' letter, while Agustí also highlighted the Cueva Victoria incident, which was closer in time.[98] It seemed then that all of these episodes accumulated and finally resulted in the *Diputació*'s decision.[99] What is clear is that Gibert's media presence had a great deal of significance in this choice.

93 For instance: A. M. M., *ABC* 1985.

94 "división del equipo", Vilardebó, *ABC* 1985 and Fernández, *El Ideal de Granada* 1985.

95 "permiso verbal", "no debe demorarse por razones puramente administrativas", Jover, *La Vanguardia* 1985.

96 Interview with Agustí 2012 and Interview with Marcet 2013.

97 "No volien escàndols", Interview with Marcet 2013.

98 Gibert 2004, 56 and Interview with Agustí 2012.

99 Interview with Marcet 2013.

At the same time, Agustí's appointment also showed that his background position had benefited him. Despite being one of the members of the Orce Man discovery team, when a new director was required he was a valuable candidate since his image had not been damaged by the controversy.

To sum up, the Cueva Victoria finding, used by Gibert as a 'tool' to reinforce his claims regarding the Orce Man, had turned against him.[100] The similar patterns of the public appearance of the phalanx finding and the public presentation of the Orce Man discovery did not go unnoticed by the story's historical actors, who feared that history might repeat itself. The Murcia politicians did not want a controversy over palaeontological or archaeological findings to affect their political image as seemed to have happened to the Andalusian politicians. And the three professors did not want more 'negative' fuss about palaeontology in Spain and made clear their position against Gibert's claims and the way he presented them to the public. Moreover, they used their authority as university professors to judge in public Gibert's scientific expertise. As Brian Martin argued in the fluoridation controversy, this example shows us the mechanisms by which scientific "orthodoxy is established and perpetuated". For Martin, it is the "persistence of a minority opposition, which ensures that the exercise of power in science is, to some extent, brought out into the open."[101] Like the *El País* front page after the de Lumleys' phone call, the letter from the three professors shows how Gibert's public resistance to accepting powerful scientists' verdicts brought that very power into the media, in this case Spanish newspapers, which became the means used to reassure and reinforce their dominant position in science.

However, Jaume Truyols (one of the three professors) has already appeared in this book. He was privately praising the discoverers and the "amazing" Orce Man, which he had learned about in newspapers and which he claimed should appear in all "human palaeontology textbooks".[102] At that time, the finding was a great success widely reported on in the media, had just been published in the *Institut's* journal, and its inner part was still not visible. Almost two years later, this same public presentation of the Orce bone was considered "hasty" and "unscientific".[103] What Truyols considered 'unscientific' was not Gibert's public dissemination of the findings before proper studies, especially when these

100 Martin's concept of 'tool' in Martin 1991, 8 and 156.

101 The fluoridation controversy arose from concerns over the fluoridation of public water supplies. Martin 1991, 8 tried to develop a 'power picture of science' in his analysis of this controversy.

102 AMC-ICPMC: Letter from Jaume Truyols to Miquel Crusafont, 25 June 1983.

103 De Renzi/Porta/Truyols, *Avui* 1985.

findings were as relevant as the Orce Man supposed to be, but his insistence on bringing scientific dispute and controversy into the Spanish public sphere (despite the fact that it was not Gibert that made the controversy public in the first place).[104] It was therefore Gibert's willingness to fight against powerful scientists' statements in public that led Truyols and the others to 'defend' palaeontology, and not the hastiness of the public presentation of discoveries that were, indeed, unproblematic when there was no controversy.

The three professors wanted to draw a line between what they considered science and what they did not, placing Gibert outside of the former. For them, scientific doubts and controversies must be removed from science, making this a clear performance of the so-called 'expulsion' type of 'boundary work'.[105] This time the boundary had to be drawn not between those who popularise and those who do not, or between those who believed that the Orce fragment was a hominid and those who did not, but between those who wanted to keep the public controversy alive (mainly Gibert) and those who did not. Moreover, and as Gieryn states referring to the Cold Fusion case, the boundary not only separated 'controversial media science' and 'cautious established science' but also determined credibility among scientists. The three professors maintained their credibility. Gibert was, from then on, a "mere fossil collector".[106]

2.3 Science in a 'Different Dimension'

After Jordi Agustí's appointment as director of the *Institut de Paleontologia de Sabadell*, the conflict between the discoverers of the Orce bone became more apparent. It is clear though that the supposed 'credibility damage' caused by the Orce controversy was completely absorbed by Gibert, while Agustí and Moyà-Solà seemed not to be affected by it (as Agustí's appointment as director suggests). What mainly caused this 'damage' was not the *El País* front page, but Gibert's way of defending himself, the Cueva Victoria finding, and the reactions to both. In any case, the split in the discoverers' team was complete by the beginning of 1986. With Agustí in charge, the official position of the *Institut* became that of denying hominid presence in Venta Micena.[107] For Gibert, who

104 Goodell 1977, 92, makes a similar point by saying that keeping controversy out of the public eye is one of the unwritten laws of science; laws that visible scientists, like Gibert, often break.

105 Gieryn 1999, 15–16.

106 *Ibid.*, 223–224.

107 Personal Archive Jordi Agustí: Letter from Gibert to Josep Maria Huguet, 21 May 1986.

kept his position in the institution, this meant that his own projects were being ignored by the *Institut*'s 'official' lines of research.[108]

Gibert then tried to reinforce his position by sending letters to politicians and scientists that could influence the *Institut*'s development. He sent a letter to Jordi Labòria, head of the *Diputació*'s Department of Culture, in which requested "the creation of a new line of research that would amend this error" (the exclusion of his projects from the lines of research).[109] Likewise, Gibert sent letters to the members of the *Institut*'s *Scientific Advisory Board*, created to advise the *Institut* on scientific issues.[110] Gibert wanted their support for his own research project, which, according to him, was one of the projects that aroused the most "social interest" and "expectations among the scientific community".[111]

At the end of February 1986, the *Junta de Andalucía*'s Department of Culture sent a letter to Gibert announcing that the excavation permit for that summer in Venta Micena was denied to his team.[112] This was the first time that this permit was denied to Gibert or the *Institut de Paleontologia*. The official reason given for denying the permit was that the previous year (1985) the excavation report was incorrect and the remains recovered were not "properly classified".[113] For Gibert, there was one other reason: doubts around the human classification of the Orce Man and therefore the hominid presence in Venta Micena had spread among the *Junta* politicians.[114] Once again, Gibert sent letters to members of the *Institut*'s *Scientific Advisory Board* asking for support to get the permit back.[115] In the press, the causes of the denial were described as "exclusively scientific".[116]

It seemed that there was yet another reason for the denial: doubts about Gibert as a scientist. On 1 July 1986, Gibert, Agustí, the professors Truyols and

108 *Ibid.* and AJG-ICPMC: Minutes from the meeting of the *Scientific Advisory Board* of the *Institut de Paleontologia Miquel Crusafont*, 1 July 1986.

109 "la creación de una nueva línea de investigación que enmiende este error", AJG-ICPMC: Letter from Gibert to Jordi Labòria, 21 May 1986.

110 For instance, Personal Archive Jordi Agustí: Letter from Josep Gibert to Josep Maria Huguet, 21 May 1986.

111 "interés social", "expectaciones en la comunidad científica", Personal Archive Jordi Agustí: Letter from Josep Gibert to Josep Maria Huguet, 21 May 1986.

112 Personal Archive Jordi Agustí: Letter from the Junta de Andalucía to Josep Gibert, 27 February 1986.

113 "classificados debidamente" *Ibid.*

114 AJG-ICPMC: Letter from Josep Gibert to Jacint Nadal, 6 June 1986.

115 *Ibid.* and Personal Archive Jordi Agustí: Letter from Gibert to Josep Maria Huguet, 6 June 1986.

116 "exclusivamente científicas", González-Besada, *ABC* 1986.

de Porta, and the *Diputació* politicians Jordi Labòria and Roger Marcet, among others, attended a meeting of the *Institut*'s *Scientific Advisory Board*, the minutes of which have been preserved.[117] During the meeting, Gibert informed the Board that the excavation permit for Venta Micena had been denied by the *Junta de Andalucía*. Jordi Labòria stated that contact with the *Junta* had already been made and that according to them there was no "institutional issue" with the *Diputació* or the *Institut*, but that the problem was "rather personal".[118] The 'boundary-work' was apparently working and Gibert was beginning to be seen by those who granted the permits as an unserious scientist with no support from the palaeontological community.

In May 1986, the Spanish popular science magazine *Muy Interesante* published an interview with a palaeontologist from Madrid, María Teresa Alberdi, who at that time was working on the geology and palaeontology of the Orce area. The interviewer, after some general questions about palaeontology, asked for her opinion on the Orce bone. Alberdi answered that it could be human but it also could be "a bluff". Moreover, for her, the discoverers should have studied the fragment in detail before "arranging the show".[119] Alberdi also added that she "has not seen the bone, nor has Emiliano Aguirre, number one Spanish palaeoanthropologist [...]", but that she knew what she knew about it because she "had seen it in the newspapers".[120] Gibert took Alberdi's comments as a personal attack and sent a letter to the director of *Muy Interesante* in which he defended himself stating that "reliable results" on the Orce Man had been reached and would be presented at a conference in Southampton the following September.[121] Gibert further claimed that Alberdi's interview revealed the "deplorable level" of scientific discussion in Spain which led to "comments, criticisms, and lies spread behind my back". Gibert also looked back on the history of human palaeontology, which is "full of situations like mine".[122] For him, important discoveries had always been controversial. These were the first stirrings of a discourse that included the Orce

117 AJG-ICPMC: Minutes from the meeting of the *Scientific Advisory Board* of the *Institut de Paleontologia Miquel Crusafont*, 1 July 1986.

118 "Problema institucional", "més aviat personal", *Ibid.*

119 "Un bluff", "organizar el show", Torreiglesias 1986.

120 "no he visto el hueso, tampoco Emiliano Aguirre, número uno de la Paleontología humana española", "pero según lo que he visto en los periódicos ...", Torreiglesias 1986.

121 "resultados fiables", Personal Archive Jordi Agustí: Letter from Josep Gibert to the director of *Muy Interesante*, 3 June 1986.

122 "nivel lamentable", "comentarios, críticas y chismes realizados a espaldas mías", "llena de situaciones como la mía", *Ibid.*

case among controversies in palaeoanthropology and that would become common in Gibert's career.

In the summer of 1986, the forthcoming international conference in Southampton became widely famous due to a dispute regarding the presence or not of the South African delegates as a protest against apartheid. After a widely reported debate, a new body, the World Archaeological Congress, emerged and finally organised the conference in Southampton.[123] But this dispute received few comments in the Spanish press, which was more interested in the Orce Man's presence at the event.[124] Gibert portrayed the conference as the place where "doubts will come to an end" and their conclusions about the Orce Man would be "confirmed" by the international community. He hoped "that the truth about the human origin" of the Orce fragment would be "ratified" in the Southampton conference.[125] Characters such as Richard Leakey, who also gave a talk, were called upon by the press as the experts that would give "definitive backing" to Gibert and his discoveries.[126] Moreover, international support would "prompt, without a doubt, the continuation of the excavations at Venta Micena."[127] According to *ABC*, in Andalusia, some archaeologists had argued that perhaps the *Junta* was waiting to see the experts' opinions from Southampton in order to take a decision on the Venta Micena excavation.[128] Newspapers highlighted that the results of an "immunological study" that had been performed on the Orce bone would also be presented at the conference. According to the press, this study, which was performed "for the first time in Europe" in the *Universidad de Granada*, found proteins in the Orce Man "that would confirm" the human nature of the bone.[129]

On 20 August 1986, Gibert, who considered himself as having the reputation of an isolated scientist, was declared adopted son of Orce due to his

123 For a more detailed history of the Southampton conference and the creation of the World Archaeological Congress see: Golson 1986 and Ucko 1987.

124 One of the few comments on the Aparheid dispute, which also commented Gibert's presence in the Conference: Ap y EFE, *La Vanguardia* 1986.

125 "dudas se acabarán", "confirmado", "espera que la verdad sobre el origen humano", "ratificada", Paredes, *La Vanguardia* 1986. See too: Vilardebó, *ABC* 1986a; Vilardebó, *ABC* 1986b; MacPherson, *El Periódico* 1986; González-Besada, *ABC* 1986; García, *El País* 1986; and J. A., *Diari de Sabadell* 1986c.

126 "espaldarazo definitivo", González-Besada, *ABC* 1986.

127 "induciría, sin duda alguna, a la continuación de las excavaciones en Venta Micena", González-Besada, *ABC* 1986.

128 González-Besada, *ABC* 1986.

129 "por primera vez en Europa", "que confirmaria", Vilardebó, *ABC* 1986 and García, *El País* 1986.

"promotion and defence of the cultural values of the town and for his efforts to keep the majority of the remains in Orce".[130] Gibert also stated that he was grateful to Orce's inhabitants for this honour as they always helped him work comfortably, and he knew how hard it was "to work in a hostile environment" (in a clear reference to what was happening at the *Institut*).[131] The fact that Gibert portrayed himself as an 'isolated scientist' working in a harsh environment highlights the tendency of some visible scientists to portray themselves as ignored outcasts, thus positioning themselves against the 'establishment'.[132] This position leads to more media attention, creating a 'loop' that causes the perpetuation of the controversy.[133] The same day that he left for Southampton, Gibert gave a lengthy interview to the Sabadell local newspaper *Diari de Sabadell*. The journalist portrayed him as a "cursed scientist" but who, after being "ridiculed" in academic circles and in the media (using the *El Papus* issue as an example), had "recovered public dignity".[134] When asked, Gibert said he preferred the term "controversial scientist" and stated that after the Southampton conference and a meeting in Sabadell, he would "convince those who still doubt me".[135] Finally, Gibert quoted Thomas S. Kuhn's *The Structure of Scientific Revolutions* to state that in science "all changes meet with resistance and controversy".[136] Kuhn and his notion of paradigm shift would also be a recurring part of Gibert's discourse to legitimate his position and to keep up the controversy.[137]

In September, the Southampton conference was reported on almost daily in the Spanish press with headlines like 'The Orce cranium is human, as confirmed by the World Archaeological Congress'.[138] It was also highlighted that "Gibert was not challenged", and therefore the evidence "settle[d] the controversy".[139] According to Gibert, the conference attendees pointed out that the

130 "Gibert se considera a él mismo un científico con fama de científico aislado", García, *El País* 1986. "... la labor de promoción y defensa de los valores culturales de esta localidad, y por su trabajo en favor de la permanencia en Orce de la mayor parte de los vestigios encontrados", Vilardebó, *ABC* 1986b.

131 "trabajar en un medio hostil", MacPherson, *El Periódico* 1986.

132 For more on this see: Goodell 1977, 23–24 and 180; Goodell 1987; and Fahnestock 1989.

133 Weingart 1998, 871–872.

134 "científico maldito", "ridiculizado", "recobrado la dignidad pública", J. A., *Diari de Sabadell* 1986c.

135 "científico polémico", "convencer a quienes todavía dudan de mi", *Ibid.*

136 "todo cambio es recibido con resistencia y polémica", *Ibid.*

137 See section 4.4.

138 'El cráneo de Orce es humano, según confirma el Congreso Arqueológico Mundial', Moreno, *El País* 1986.

139 "Gibert no fue discutido", "zanjar la polémica", Piñol, *La Vanguardia* 1986.

immunological studies were "decisive" and that the human characteristics of the bone had been "absolutely proven", so from then on it would be "very hard" to contest the evidence.[140]

To sum up, before the conference, the international meeting's verdict was seen in the Spanish public sphere and among scientists in Andalusia as 'definitive'. Afterwards, the apparent positive response of the experts seemed to confirm Gibert's position regarding the bone. The conference was also the first place where the immunological studies of the bone were presented to the scientific community. The following December 1986, *El Periódico* published a two-page report on the Orce research which highlighted the international scientists' support of the Orce Man. The immunological research carried out on the Orce bone by both the Spanish research team and Lowenstein research in America, together with Domènec Campillo's anatomical study were presented as the final validation of Gibert's position regarding the bone.[141] Here it is worth noting the contradiction between Gibert's self-proclaimed isolation and the international support he apparently received, a contradiction that will be explored further later.[142]

Later, in November 1986, the *Institut de Paleontologia* hosted a tribute to Miquel Crusafont with a three-day meeting with almost 150 delegates, which was reported on in local and national newspapers. The first session was dedicated to remembering Crusafont. Jaume Truyols (Crusafont's scientific heir), Emiliano Aguirre (one of the most prestigious Spanish palaeontologists and also Crusafont's friend), Jordi Agustí (director of the *Institut*), and the *Diputació* politician Jordi Labòria all spoke during the ceremony.[143] The following two days were devoted to scientific communications. Among them, only one was commented on in newspapers: Gibert's 'defence' of the Orce Man.[144] In the *Diari de Sabadell*, the reactions among those who attended Gibert's presentation were described as being between "expectation" and "scepticism" (figure 10).[145] The local newspaper then went on to refer to Emiliano Aguirre's remarks on the Orce Man. Aguirre stated that he was "pleased" to see how the discovery had motivated several very important studies in Spanish palaeontology. Then he added that he wished "that new discoveries, new fossils, would confirm the

140 "Determinantes, absolutamente probadas", Moreno, *El País* 1986; "Lo tendrán muy difícil", Piñol, *La Vanguardia* 1986. See also: Redacción, *El Periódico* 1986b and Redacción, *El Periódico* 1986c.

141 Barata, *El Periódico* 1986.

142 This contradiction is fairly common in other palaeoanthropologist public discourse, see Hochadel 2013a.

143 J. A., *Diari de Sabadell* 1986a and Mercade, *La Vanguardia* 1986a.

144 Mercade, *La Vanguardia* 1986b; J. A., *Diari de Sabadell* 1986b; Redacción, *El Periódico* 1986d.

145 "expectación", "escepticismo", J. A., *Diari de Sabadell* 1986b.

Clausura de las Segundas Jornadas de Paleontología

La presentación del «Hombre de Orce» acogida con escepticismo

La comunicación del doctor Josep Gibert sobre el «Hombre de Orce» levantó expectación ayer tarde en las Segundas Jornadas de Paleontología. La polémica sigue viva, tal como les contamos en la **página 3.**

FIGURE 10 Gibert's talk on the front page of the *Diari de Sabadell*
FROM J. A., *DIARI DE SABADELL* 1986B

thought-provoking hypotheses that Dr. Josep Gibert has shown us."[146] According to the newspaper, these vague statements neither for or against Gibert's interpretation of the bone were seen among some of the attendees as a "slight disavowal" of Gibert's work.[147] *El Periódico* even stated that attendees "questioned" the Orce Man and criticised Gibert and his research.[148] Yet, for Gibert, who was quoted in the *Diari de Sabadell* piece, Aguirre's words were a "real, although subtle, vote of confidence" for his research.[149]

146 "satisfecho", "deseo que nuevos hallazgos, nuevos fósiles, vengan a confirmar las sugestivas hipótesis que el Dr. Gibert nos ha mostrado", *Ibid.*
147 "leve desautorización", *Ibid.*
148 Redacción, *El Periódico* 1986d.
149 "un auténtico, aunque taimado, voto de confianza", J. A., *Diari de Sabadell* 1986b.

The analysis of these meetings reveals three important points. Firstly, both the Southampton conference and the Sabadell meeting were actively followed by several newspapers, thus transforming a closed scientific presentation for experts into a public scientific event, the content of and reactions to which people could to some extent follow.[150] Secondly, the 'expectation' about both of Gibert's presentations shows how his research was still considered important and worth following, not isolated or marginalised. Thirdly, the opposite reaction to Aguirre's statements would be a constant during the Orce controversy. Gibert and his critics would use the same statements by renowned scientists, who were usually neutral and polite, in completely different ways.

To sum up, Gibert sought support, communicated his ideas, and 'defended his honour' through an array of correspondents, from the members of the *Institut's Scientific Advisory Board* to the director of a magazine. He strove to get scientists to support his research and politicians to give him excavation permits, and also tried to 'clean' his own public credibility and image whenever they were questioned. Publicly, Gibert wanted to control his own scientific status and at the same time control the Orce Man's public appearances. During this period, little by little, he shaped a 'discourse' that would be further developed during the following years. The main lines of this discourse already appeared in these early years of the controversy: an 'isolated' scientist who changes the 'scientific paradigm' (similar to many other controversies in human palaeontology) and who fights against a mainstream that is not 'scientific enough' due to how 'young' Spanish science was at the time, which, for Gibert, did not allow for proper scientific debate.[151]

Finally, Gibert's presence at the Southampton conference and in Sabadell was covered by the press, including reactions, developments, and aftermath. Newspapers wanting to know more about the Orce controversy showed the public how a scientific conference worked, turning it into a public event, and at the same time turning the newspaper into a scientific forum. Spanish scientists, like Alberdi, found out about the Orce Man in newspapers and laypeople found out about a scientific conference in just the same way. As Stephen Hilgartner put it, "popularisation [and science] is a category not necessarily established by its genre, but also by the context of reception."[152]

150 More on this role of scientific conferences in Chapter 3 and in Carandell 2015a.

151 More on Gibert's comments on the state of Spanish science: Interview with Josep Gibert 2003.

152 Hilgartner 1990, 529. See also: Gregory 2003, 26.

After the positive reception of Gibert's work in Southampton and the not so negative one in Sabadell, and its crucial coverage in the press, in 1987, the *Junta de Andalucía* granted him an excavation permit.[153] According to Gibert, during this excavation he found the first stone tools of the Orce region in a site near Venta Micena called Barranco León.[154] During that summer, Orce's small museum, which was inaugurated after the original discovery's presentation, was improved with the addition of two new rooms thanks to the support of Gibert's team.[155] The little Orce museum was beginning to be yet another sympathetic forum to support Gibert's claims.

Nevertheless, throughout the year, the Orce controversy hardly appeared at all in the Spanish press. At the beginning of October, Gibert went to Turin to present his work at an international conference. In the opening session, which Gibert did not attend, Marie-Antoinette de Lumley criticised the Orce Man. Yet when the Spanish researcher presented his work, not one of the approximately four hundred attendees asked questions or discussed the presentation.[156] According to him, since Marie-Antoinette had nothing to ask or any criticism to make, then nobody else did either.[157]

While Gibert was in Turin, on 2 October, Jordi Agustí and Salvador Moyà-Solà (co-discoverers of the Orce Man and former collaborators of Gibert's) announced to the press the publication of an article in the Spanish scientific journal *Estudios Geológicos* in which they argued that the Orce cranial fragment belonged to an equine.[158] The article was received by the journal on 20 October and not accepted until 21 November, so the announcement to the press was made not only before its acceptance but even before they had sent it to the journal.[159] The *Institut* researchers had to establish their position first in the public arena and only afterwards in the scientific forum. It is also worth noting that, according to Agustí, they asked the *Diputació* politicians if they could "distance themselves" from the Orce Man issue, a thing that they had not yet done publicly.[160] Agustí stated in newspapers: "the debate should have remained in the scientific publications, but the journalists' treatment of the

153 Gibert 2004, 79.

154 *Ibid.*, 83.

155 Martínez-Navarro 1993, 45.

156 Barata, *El Periódico* 1987b and Gibert 2004, 77–78.

157 Gibert 2004, 78.

158 Mercadé, *La Vanguardia* 1987; Redacción, *El País* 1987; Barata, *El Periódico* 1987a. The scientific article is Agustí/Moyà-Solà 1987.

159 Agustí/Moyà-Solà 1987.

160 "Desmarcarse", Interview with Agustí 2012.

issue has given it a different dimension."[161] In the scientific article, Moyà-Solà and Agustí also complained that Gibert had not allowed them to take an x-ray of the fragment.[162] That same day, both *El País* and *La Vanguardia* phoned Gibert in Turin to ask him for his opinion on Agustí and Moyà-Solà's statements. Gibert said that this "attack" happened due to "extra-scientific issues"; they wanted to undermine the significance of "his" sites.[163]

Therefore, after having excavation permit on 1987, which gave Gibert momentary control over the site and its discourses and claims, bringing an apparent end to the early discussion; the de Lumleys, Agustí and Moyà-Solà brought the debate back to the public arena. After this new controversy, in 1988, the Venta Micena excavation permit was once more denied to Gibert's team.[164] In any case, the way that Agustí and Moyà-Solà made their aggressive move against Gibert shows how in order to clearly state and lay down their opinion regarding the Orce fragment, the first, fundamental, and most important step was the public statement.[165] This way of acting occurs often in public and political scientific controversies or important discoveries. The most well-known case, analysed several times by sociologists and historians of science, is the Cold Fusion saga.[166] Yet, it is also common in palaeoanthropological debates like the *Homo floresiensis* dispute or the world-wide presentation of Ida, 'the link' (not 'missing' anymore).[167] These studies have highlighted how the public exposure of scientific research can have several interests and objectives behind it.

What were Agustí and Moyà-Solà's aims? If one of the critiques of Gibert was his penchant for excess public exposure, then why did they jeopardise their morally superior position of having remained silent? Why the need to establish their new scientific position publicly? When this move was made, Agustí was already director of the *Institut*, so the reasons were not about accessing a better professional position. Agustí and Moyà-Solà in fact, decided that it was necessary to ask to the *Diputació* politicians if they could do this

161 "El Debate debería haberse mantenido en los cauces de las publicaciones científicas pero que el tratamiento periodístico del tema le ha dado una dimensión distinta", Mercadé, *La Vanguardia* 1987.

162 Agustí/Moyà-Solà 1987, 536.

163 "Motivos extracientíficos", Mercadé, *La Vanguardia* 1987 and Redacción, *El País* 1987.

164 Martínez-Navarro 1993, 26.

165 A similar instance regarding the cold fusion controversy can be found in Gieryn 1999, 187.

166 For instance: Lewenstein 1995 and Bucchi 1998, 36–81.

167 For the Flores 'hobbit' see Goulden 2013, and for the Ida public presentation see Kjærgaard 2011a.

move. It was not an easy move. Why they did it then? What were the motivations for exposing themselves this way?

According to Agustí and Moyà-Solà, they did it because Gibert was publicly saying that "everybody in the *Institut* thinks that the bone is from a hominid", and this was not true.[168] Yet, after Southampton and Sabadell's presentations and the 1987 excavation permit, Gibert was achieving an stronger public position and his claims were appearing more often in the media, so, from Agustí's and Moyà-Solà's perspective, Gibert had to be contested in public. Moreover, with their new position, Agustí and Moyà-Solà also tried to distance themselves once and for all from the Orce Man controversy. Beyond their own opinion on the fragment, they did not want to be linked to Gibert's controversial position and tactics. Their desire to be distanced from Gibert's opinion regarding the bone was also a desire to be removed from public dispute, from public exposure to controversy. They wanted to introduce themselves into uncontroversial 'proper science', what Truyols considered science. This delimitation not only separated different scientific positions but also defined credibility among scientists. Agustí and Moyà-Solà wanted to be considered credible, and to do so they had to go through the painful process of briefly exposing their position publicly.

The Orce story shows how the media presence of science and scientific controversy matters to politicians; for instance, in their complaints about the 'excessive' Cueva Victoria dissemination efforts or in the *Diputació* politicians statements that they 'did not want scandals'. Politicians declared that their decisions were based on purely 'scientific' reasons, but the media also had an impact on their actions. The positive reaction of the Junta to the media coverage of the Southampton conference or the permit denial after Agustí and Moyà-Solà's announcement reveal how politicians made decisions in relation to scientific issues taking into account what the press said about them. The way that the controversy was presented in the media lobbied the politicians in their decision-making. These decisions at the same time affected the way that scientific research was taking place. So, the Orce Man controversy sheds light on how media coverage influences science through politicians.[169] This, presumably, is a phenomenon that could happen often but that such a highly public controversy like the Orce Man makes especially visible. In addition, the process that turned Gibert into a 'marginalised scientist' was not something

168 "tothom a l'Institut pensa que l'òs és un homínid", Interview with Agustí 2012 and Interview with Moyà-Solà 2012. Also Personal Archive Jordi Agustí: Letter from Moyà-Solà to unknown, 24 September 1987.

169 See another instance in González-Silva 2007.

that happened all of a sudden after the *El País* front page or after the *El Papus* cover (figures 8 and 9). It was, instead, a complex process that began with the early popularisation effort and little by little accumulated accusations, 'boundary-work', but also self-portrayals, until it became a well-established notion in Gibert's own discourse, other scientists' views, and even in journalists' considerations. Somehow then, beyond the denial of excavation permits, Gibert's scientific isolation was in many ways a self-made position, a chosen one that generated benefits to him, but that at the same time, also drawbacks and contradictions.

Finally, for a little over three years, from May 1984 (when *El País* published the de Lumley reclassification) to October 1987 (when announced the publication of the first scientific paper dealing with this statement), the scientific debate around the Orce Man took place mainly in the public sphere. The Southampton conference or the Sabadell meeting, where 'scientific' discussions were held, were widely reported on in the media. Journalists, politicians, and scientists gave their opinion to the press, which became the main communication channel for the controversy. Moreover, Moyà-Solà and Agustí only published their article because Gibert insisted on saying that everybody in the *Institut* agreed that the Orce Man was in fact a man. Had Gibert not insisted on the hominid argument, had he forgotten about the controversy, the *El País* front page could perhaps have been enough to establish that the Orce Man, the famous 'First European', was in fact an equine. Thus, this shows us how in this instance the *El País* newspaper acted as a scientific channel among the community. Equally, if any scientist wanted to keep up to date on research on the Orce fragment and the dispute about it, they could attend any scientific conference that Gibert participated in, but by reading newspapers and other public media they could not only get information on these conferences, but also several details on the bone and Gibert's and his opponents' moves, projects, research results, and opinions. Therefore, newspapers and the mass media acted as 'scientific channels' for the scientific community to get information on Orce. The way that Agustí and Moyà-Solà announced their publication clearly shows how traditional scientific publications were not enough to disseminate their opinion among peers, and a public statement was also necessary to convince them. At least during these first three years (from 1984 to 1987), the press and the mass media in general were spaces were science was legitimated and discussed beyond traditional scientific conferences or journals.[170] The

170 Shinn/Cloître 1985 and Bucchi 1998, use the concept of 'crystallization' or ' temporary crystallization', according to which, certain concepts or discoveries fit better in certain

media visibility of the Orce issue had authority in the validation of both sides claims.[171] In this supposed 'different dimension', that Agustí talked about, the media became a crucial step in the process of presentation, discussion and final validation of scientific knowledge.[172]

media but not another. Somehow the Orce Man issue was 'crystallized' in the scientific level and, as Bucchi states, 'require(d) the intervention of the public to determine the success of one party over another', Bucchi 1998, 11.

171 Another example in Van Dijck 2008, 390.

172 See a similar point in Gieryn 1999, 200–201.

CHAPTER 3

Conference (1987–1996)

It is a warm September evening in the central square of a small Andalusian village. Several screens and a number of chairs have been set up. People take their seats. Yet, instead of the latest Hollywood movie, they are about to watch the final session of an international scientific conference on human palaeontology. The session begins. At some point during the live broadcast, the floor is given to a well-known international scholar. "He's a prominent authority" can be heard in the square. The scholar begins: "When I came, some days ago, I was very sceptical about the claims made regarding this region. Now, after I have seen the sites, examined the remains, and heard the presentations, I am fully convinced of the validity of these claims." The crowd in the square is excited; it seemed like all of the doubts were fading away.

After years of recovery, due to very specific public and political conditions, in September 1995, Gibert achieved to organise a major international conference. This conference was used by him as a reflection of his own research project and therefore of his own credibility as a scientist. For him, its organisation validated his own scientific claims. This reflection was intended to be seen not only by scientists but also by journalists and politicians. Its final success seemed to ratify Gibert's scientific, political, and public positions.

3.1 Gibert's Research Team and the Conference Preparation

In the 1990s, the hunt for the First European was still alive among palaeoanthropologists. Several new claims for the oldest European remains appeared and some of the old ones were recovered. In 1989, the French scientist Eugène Bonifay organised a conference on this topic funded by the scientific affairs division of NATO, the European Council, France's scientific research council (CNRS), and its *Ministère de Culture*. With an article on the Orce remains by Gibert and his team, the proceedings of this conference were published in 1991 under the title *Les Premiers Européens*. Bonifay stated that the 1 million-year-old human presence in Europe was easy for scholars to accept but that older remains were always controversial. Bonifay himself had been claiming a very old first settlement in Europe, up to 2 million years old. From 1991, several hominids remains controversially dated around 1.7 million years old were found in

 DOI:10.1163/9789004431508_005

the Georgian town of Dmanisi. What later became the *Homo georgicus* was again a very serious contender for the First European.[1]

In 1993, a workshop on the 'Earliest Occupation of Europe' was held in Tautavel, funded by the European Science Foundation, this time without Gibert or any of his collaborators participation. In the proceedings of this workshop, edited by Wil Roebroeks and Thijs van Kolfshoten, and later in an article in *Antiquity*, these two archaeologists established a 'short chronology' stating that the evidence on hominid remains in Europe before 0.5 million years ago was very weak.[2] In 1996, Robin Dennell and Roebroeks again published an article in which they 'revisited' the short chronology after Atapuerca's and Orce's findings. They stated that southern Europe could have been colonised before 0.5 million years ago but always sporadically, with no 'continuous' occupation.[3] Finally, the well-known French palaeoanthropologist Yves Coppens, one of the co-discoverers of Lucy, was another of the main proponents (together with Bonifay) of an alternative 'long chronology', stating that hominids could have been present in Europe around 2 or 2.5 million years ago.[4]

From September 1995 to September 1997, the *National Geographic* published a series of articles called 'The Dawn of Humans'. Within this series, in July 1997, the magazine published an article on 'The First Europeans'.[5] The old Mauer mandible and the Tautavel Man were there, but new contenders also joined the race. The Ceprano Man was a skullcap found in central Italy, then thought to be around 0.8 million years old.[6] In England, the Boxgrove tibia was dated at around 0.5 million years old and headlined in *Nature* with the title 'First European?'[7] In Atapuerca, northern Spain, a new species of hominid fossil was dated at almost 0.8 million years old.[8] Among them all, there was also a 'controversial contender': the Venta Micena remains, placed in this article at between 1.2 and 1 million years old.[9] The Orce Man cranial fragment was still its main flagship and occupied a central position in the photo that illustrated the

1 For the conference proceedings see: Bonifay/Vandermeersch 1991. A compilation of Bonifay's claims in: Bonifay 2002. Bonifay also supported Gibert's claims in a documentary: Navarro 2010. For Georgia Agustí/Lordkipanidze 2005.

2 Roebroeks/Van Kolfschoten 1994; Roebroeks/Van Kolfschoten 1995a; and Roebroeks/Van Kolfschoten 1995b.

3 Dennell/Roebroeks 1996, 540.

4 Coppens 2005, 50–52.

5 Gore 1997.

6 Ascenzi et al. 1996 and Nomade et al. 2011.

7 Roberts/Stringer/Parfitt 1994 and Gamble 1994.

8 See, for instance: Carbonell et al. 1995 and also Hochadel 2013b, 73–112.

9 Gore 1997, 100.

article. During these years, a theoretical rivalry between the two Spanish sites (Atapuerca and Orce) became more and more apparent in the media, and the Atapuerca leaders were one of the main voices against the validity of Gibert's claims.[10]

In this context, Gibert and his team organised a major international conference in Orce in 1995 where the study of hominid arrival to Europe was central. Again, it seemed that Gibert's claims were not unique and isolated but part of a wider debate regarding the settlement of hominids in Europe.

∴

The Orce Man saw less media attention during the years following October 1987, when Moyà-Solà and Agustí appeared in the newspapers announcing that it was for them a member of the *Equus* genus. Gibert did not get an excavation permit for Venta Micena for 1988 or 1989, but instead developed an in-depth study of the bone fragment and promoted the development of the *Orce Museum*.[11] During what journalists called Gibert's 'great recovery' from the public exposure of the controversy, he devoted himself to set up a research team around him, to responding to Moyà-Solà and Agustí's article, attending conferences, giving formal and popular talks, and publishing all the research in self-published monographs, conference proceedings, and scientific journals.[12]

During those years, Gibert gathered an interdisciplinary team with varying degrees of collaboration and involvement. Domènec Campillo, an eminent palaeopathologist and neurosurgeon who, although he did not have any official link with Gibert, collaborated with him in the anatomical study of the Orce Man. Since his first diagnosis just after the discovery he never changed his opinion.[13] He edited and published several papers and monographs with Gibert and in 2002 published a book on the famous cranial fragment.[14] Enrique García-Olivares was an immunologist from the *Universidad de Granada*. Back in 1986, he got in contact with Gibert, who suggested him to conduct

10 See Hochadel 2013b, 188–192.

11 Martínez-Navarro 1993, 45.

12 'Gran recuperación', Fontova, *El Periódico* 1990. A response to Agustí and Moyà-Solà in the same journal by Gibert et al. 1989a. For scientific conferences: Gibert 2004, 79–83. For popular talks, for instance: Redacción, *La Vanguardia* 1988; Redacción, *La Vanguardia* 1989. For monographs: Gibert/Campillo/García-Olivares (eds.) 1989; Gibert et al. (eds.) 1993. For other articles: Gibert et al. 1989b; Gibert/Jimencz 1991; Gibert et al. 1991; and Palmqvist/Gibert/Martínez-Navarro 1992.

13 Campillo 2002 and Interview with Campillo 2011.

14 Gibert/Campillo/García-Olivares (eds.) 1989, Gibert et al. (eds.) 1993, and Campillo 2002.

an immunological test on the Orce remain to look for human albumin.[15] This technique was applied for the first time in Europe to the Orce Man, after it had been developed in the United States by Jerold Lowenstein.[16] Later, Lowenstein himself also applied this method to the Orce remains.[17] Both laboratories' results were positive for Gibert's interests; human albumin was detected in the analyses. By the time of the Orce conference, these results had been presented at several other conferences but were only published as preliminary results in self-published monographs and as part of a PhD thesis.[18] The British archaeologist Michael J. Walker, who knew Gibert since the 1970s, had collaborated with him since 1992 in the excavations carried out at Cabezo Gordo, Murcia, where both were co-directors of the research project (see map 1). The team found several hominid remains, which, although indisputable, were also much younger than the Orce ones; around half a million years old.[19] The Cabezo Gordo site became one more form of support for Gibert's claims.

In 1992, during the installation of an electric power pole, a new site with several stone tools was found and was from then on known as Fuente Nueva 3. Sometime later, in 1994, the French archaeologist Alain Turq from the *Musée National de Préhistoire* (Les Eyzies) directed an exploratory excavation in the area. This excavation was repeated in 1995, the year of the conference. Turq, who attended the conference, published the stone tools finding in 1995 together with Gibert and others.[20] Bienvenido Martínez-Navarro (see figure 12) was one of the high school students that participated in the 1982 excavation when the Orce bone was found. He later became Gibert's doctoral student until he graduated in 1991 with a PhD thesis on Venta Micena's fauna fossils. From November 1991 he was in charge of the *Museo de Prehistoria y Paleontología* in Orce.[21] The youngest member of the group was Lluís Gibert Beotas, Gibert's son, who had only completed his undergraduate degree in geology in 1993 yet collaborated actively in the Orce conference's organisation.

Finally, last of Gibert's collaborators was Paul Palmqvist Barrena (see figure 12), a biologist and mathematician from the *Universidad de Málaga*. Just

15 Interview with García-Olivares 2013.
16 For instance: Borja et al. 1992.
17 Borja et al. 1997.
18 Conferences: Gibert et al. 1986 and Gibert/Campillo/García-Olivares 1987. PhD thesis: Borja 1995. Articles in monographs: García-Olivares et al. 1989 and Borja et al. 1992.
19 Interview with Michael Walker 2013. See also: Walker/Gibert 1992 and Walker et al. 2008.
20 Gibert 2004, 174–176 and Tixier et al. 1995.
21 PhD: Martínez-Navarro 1991. AJG-ICPMC: Employment document of Bienvenido Martínez-Navarro from the Orce Town Council, 2 December 1991 and De León-Sotelo, *ABC* 1996a.

before the conference, he published an article with Gibert based on a fractal analysis of the Orce bone in the *Journal of Human Evolution*, the most prestigious journal in the field of palaeoanthropology.[22] This paper explained the 'fractal dimensions' of the sutures of different skulls (including the controversial Orce Man) through complicated mathematical formulas that established the complexity of those sutures. In the article, Palmqvist, who had performed similar analyses on ammonites and other palaeontological remains, analysed the 'fractal dimensions' of the sagittal and lambdoid sutures of different skulls. These 'fractal dimensions' determined the degree of the sutures' design complexity.[23] Apparently, hominids showed much less complexity than equines. According to the article, this was not the first time that this technique had been applied in palaeontology or anthropology since fractal analysis had been used in studying "sutural design complexity in ammonites" and "sagittal sutures of native American skulls".[24]

Palmqvist and Gibert's article analysed the sutures of several skulls, including horses, monkeys, hominids, and present-day humans, from different historic and prehistoric excavations.[25] Among all of these specimens, there is the controversial VM-0; the Orce Man. A computer programme estimated the fractal dimensions of digitised drawings of the cranial sutures of all these specimens.[26] The results were that the Orce Man's fractal dimension value was in the range of values of infant hominids and modern humans, and also some infant monkeys, but was far from those values registered in the equine family, which showed higher complexity in suture design.[27] This result, added to the lack of monkey fossils in Venta Micena, led the authors to conclude that according to the fractal analysis, VM-0 probably belonged to a *Homo* sp. infant.[28]

At least in March 1989 and again in February 1990, Gibert and Bienvendio Martínez-Navarro travelled to Kenya to compare the Orce hominid remains with those in the *Nairobi National Museum*.[29] Apparently, Gibert and Martínez-Navarro found a similar crest to that of the Orce Man in KNM-ER 3733, an almost complete skull from the famous Koobi Fora site near the Turkana Lake, excavated by Richard Leakey. The skull was attributed to *Homo ergaster* with

22 Gibert/Palmqvist 1995.

23 Gibert/Palmqvist 1995, 571. See also Annex 1.

24 *Ibid.*, 565.

25 *Ibid.*, 567.

26 *Ibid.*, 567.

27 *Ibid.*, 572.

28 *Ibid.*, 573.

29 Personal Archive Jordi Agustí: Letter from Josep Gibert to Josep Maria Huguet, 9 January 1987. Martínez-Navarro 1993, 27–28 and Fontova, *El Periódico* 1990.

an age of 1.75 million years. This allowed Martínez-Navarro to state that the Orce crest was not as anomalous as it seemed.[30]

During this time, Gibert team also published some scientific monographs about the Orce research. The first monograph, entitled *The human remains of Orce and Cueva Victoria* and published by the *Institut de Paleontologia de Sabadell*, was publicly presented in February 1990.[31] During this presentation, Gibert announced the discovery of a new hominid remain in Venta Micena, this time an infant's humerus, found by Bienvenido Martínez-Navarro among the old collections from the site.[32] During the following days, all newspapers reported this discovery.[33] Although there is not definitive evidence, these announcements seem to be linked to the fact that later that year Gibert was granted an excavation permit for the Venta Micena site for the summer of 1990, for the first time in three years. In these 1990 excavations, a new apparently female hominid humerus was found.[34] Despite this 1990 excavation campaign, Gibert did not achieved proper excavation permits for Venta Micena for the next five years.

Three years later, in May 1993, *El País* published a report on the public presentation of the second monograph called *Orce – Cueva Victoria Project (1988–1992). Human Presence in the Lower Pleistocene in Granada and Murcia* and published by the Orce Town Council and the *Orce Museum*.[35] In this presentation, Gibert was accompanied by the prestigious French palaeoanthropologist Yves Coppens, who, according to Gibert, agreed with his theories. Gibert claimed the Orce Man had "completed his journey through the desert" since this monograph presented "irrefutable evidence" of the hominid nature of the Orce and Cueva Victoria remains (figure 11). Coppens, meanwhile, stated, according to the news piece, that there was a "strong likelihood" that the Orce Man was a hominid.[36]

---

30 Martínez-Navarro 1996.

31 'Los restos humanos de Orce y Cueva Victoria', Gibert/Campillo/García-Olivares (eds.) 1989.

32 Gibert 2004, 107. And: Sánchez et al. 1999; Gibert et al. 1999c; Gibert/Sánchez/Ribot 2002.

33 The new humerus appeared in the media together with two vertebrae and a phalanx (and even a tooth in *Diari de Sabadell*) which never appeared in the scientific literature: Fontova, *El Periódico* 1990; Redacción, *La Vanguardia* 1990; EFE, *ABC* 1990a; EFE, *ABC* 1990b; Moreno, *El País* 1990; J. A., *Diari de Sabadell* 1990.

34 Gibert 2004, 112–113.

35 *Proyecto Orce-Cueva Victoria (1988–1992). Presencia humana en el Pleistoceno inferior de Granada y Murcia*, Gibert et al. (eds.) 1993. The report of the presentation in Antón, *El País* 1993.

36 "completó su travesía del desierto", "Evidencias irrefutables", "fuerte probabilidad", Antón, *El País* 1993.

EL PAÍS, jueves 13 de mayo de 1993

El Hombre de Orce acaba su "travesía del desierto", dice su descubridor

El prestigioso paleontólogo francés Yves Coppens respalda a Josep Gibert

JACINTO ANTÓN, **Barcelona**

A los 10 años del descubrimiento de los restos fósiles de Venta Micena (Granada), el *hombre de Orce*, el más sonado —y polémico: se dijo que era un équido— hallazgo de la paleontología española, "ha concluido su travesía del desierto", según su descubridor, Josep Gibert. "Quien lea bien los datos lo tiene difícil para negar la humanidad de los restos", afirma. Gibert, que estos días presenta en varios lugares de España (ayer lo hizo en la Universidad Autónoma de Barcelona) un voluminoso libro que resume los trabajos de su equipo y presenta nuevas aportaciones, ha conseguido un aliado de excepción: el prestigioso paleontólogo francés Yves Coppens.

Para Gibert "a partir de ahora puede haber polémica, pero siempre centrada en la ciencia y ya no en la descalificación y la vehemencia".

Josep Gibert ha envejecido notablemente en los últimos años y su aspecto es hoy el de un venerable buscador de fósiles victoriano. La *travesía del desierto* (que lo ha sido tanto para su fósil como para él) ha dejado, sin embargo, huellas: un eczema nervioso en la mano derecha y una especial susceptibilidad hacia cualquier agravio. Ayer, un puñado de personas, apenas una veintena, acudieron a la presentación (en el marco de un pequeño ciclo técnico sobre las investigaciones) de la monografía *Presencia humana en el Pleistoceno inferior de Granada y Murcia (Proyecto Orce-Cueva Victoria, 1988-1992)*. Y eso que Gibert contaba con el *reclamo* de Yves Coppens, que presentó el libro —es autor del prólogo del mismo—, y dio una conferencia.

Basándose en los datos, Gibert subrayó la "demoledora evidencia" de la humanidad de siete fragmentos óseos hallados en las excavaciones (3 en Orce y 4 en Cueva Victoria), entre ellos el famoso trozo de cráneo.

También aventuró que los fósiles, "de entre 1 y 2 millones de años de antigüedad", pertenecerían, de acuerdo con las nuevas teorías que proponen nuevos tipos de primeros hombres (hasta hace relativamente poco *homo habilis* era la única denominación de los seres entre los australopitecos y el *homo erectus* —segundo hombre—), al *homo rudolfensis*. Gibert dijo que Coppens esta de acuerdo en esta hipótesis.

Coppens apoya a Gibert, pero es enormemente cauto en sus palabras. Interrogado sobre la humanidad de los restos de Orce, dijo que existe una "fuerte probabilidad" de ello. Aunque matizó: "No es incuestionable". Y añadió: "Las piezas son discutibles; no hay piezas definitivas"; y "es cierto que no hay consenso internacional". Dicho todo ello, agregó: "También es evidente que yo estoy aquí y firmo el prólogo del libro".

SÍLVIA T. COLMENERO

Yves Coppens (izquierda) y Josep Gibert, retratados ayer en Barcelona.

FIGURE 11 Gibert and Coppens after the presentation of the monograph
FROM ANTÓN, *EL PAÍS* 1993

Beyond scientific publications that were announced in newspapers, Gibert's team also did an important effort to popularize their research. A clear example came when in 1993, Martínez-Navarro published a short popularisation book on the history of the Orce Man research and its characteristics.[37] In the book, Martínez-Navarro reviewed the history of the Orce Man, put forward 'Gibert's new paradigm', and presented the work carried out at the *Museo de Paleontología* in Orce, where he was director. In 1993, Gibert signed an article in *La Vanguardia* reviewing the Orce controversy and his team's major achievements during the last few years, despite their lack of funding and excavation permits.

37 Martínez-Navarro 1993.

He also considered the great significance of the Orce Man as the 'First European'.[38] Martínez-Navarro and Gibert were trying to ensure that the public got the idea of the Orce Man that they wanted. They utilised their own story in order to control it and to achieve their aims; mainly to recover lost prestige and, as a result, get funding and support. Reviewing the history of the controversy was a way of getting followers and resources for their own stance. As sociologist Greg Myers states in his work on scientific popularisation discourses: "Being able to explain one's project and its relevance to wider society, […] is an essential part of running a large lab and getting further funding."[39]

Yet, not everything was positive for Gibert during these years. In the popularisation magazine *Revista de Arqueología,* archaeologist Raúl Monzón wrote an article harshly criticising the Catalan scientist. In his view, Gibert did not publish in scientific forums and did not give the opportunity to other colleagues to examine the remains but instead made "statements to journalists". For Monzón, these statements were made "with no more evidence (studies, analysis …) than his own words (which must have some special charm)."[40] According to him, Gibert lived off "funding that he achieved through confusion and cheap propaganda" in a scientific career that "has a lot in common with a bizarre screenplay" and nothing of what it should have: "scientific rigour and seriousness".[41] The article ended by comparing Gibert's story with a well-known fable: "who knows? […] the same thing could happen to him [Gibert] as to the lying boy with the wolf, he could really find a human remain […] and even the press might not believe him".[42] The critic to Gibert based on his tendency to go public when any discovery was made, repeated quite often among Gibert's rivals.

In January 1994, *El País* published an article entitled 'Lumley dismisses the Orce Man'.[43] It started by saying that Henry de Lumley had "with just a few words, torpedoed the slow rehabilitation of the Orce Man."[44] De Lumley was in Barcelona for a conference on the Tautavel Man, which was only dealt with at

38 Gibert, *La Vanguardia* 1993.

39 Myers 2003, 270. For more examples on funding for palaeoanthropology see: Kjærgaard 2012.

40 "declaraciones a los periodistas", "sin más pruebas (estudios, análisis, etc) que sus palabras (las cuales deben poseer un embrujo especial)", Monzón 1989. See also: Gibert 2004, 408.

41 "subvenciones que logran mediante confusionismo y propaganda barata", "tiene mucho de rocambolesco guión cinematográfico", "rigor científico y seriedad", Monzón 1989.

42 "¿quién sabe?, pudiera ocurrirle lo que al pastor mentiroso con el lobo, que realmente encuentre algún vestigio humano […] y que entonces ni la prensa le crea." *Ibid.*

43 'Lumley descalifica al Hombre de Orce', Antón, *El País* 1994.

44 "ha torpedeado con unas pocas palabras la lenta *rehabilitación* del Hombre de Orce", *Ibid.*

the end of the article. Apparently, de Lumley "dismissed" Gibert's new research and "downplayed" Coppens support of Gibert, as Coppens had "endorsed other questionable remains before."[45] The article also contained Gibert's response to de Lumley. For Gibert, if de Lumley thought that the Orce hominid was not a man, he had to "prove it scientifically".[46] Despite these attacks, Gibert has achieved a considerable amount of scientific publications, research team and public support to feel strong enough to organize a major scientific conference.

∴

The next sections will deal with the origin, organization and development of the 1995 Conference organized by Josep Gibert and his team in Orce. Yet, before that, the example of Louis Leakey's Pan-African Congress will be analysed in order to use it with comparison proposes. This example will deal with several issues that will be later analysed with more detail for the Orce case.

In January 1947, the first Pan-African Congress of Prehistory and Palaeontology was held in Nairobi, Kenya. The idea was very much Louis Leakey's "brainchild".[47] Louis was Richard's father and became perhaps the most well-known prehistorian of his time, to a large extent because he was a "great believer in popularization" and had a kind of "entrepreneurial, showbiz manner".[48] It was precisely Leakey's "constant use of newspapers" which "had further eroded his position" had given him the reputation of being a "maverick".[49] Leakey's determination to organise "a memorable meeting" transformed the 1947 congress into something that could give him "the chance to restore some of the lustre he had lost."[50] In addition, it seemed that Leakey also had the more 'scientific' goal of reinforcing his research and the general claim that Africa was the birthplace of man.[51] In Leakey's times, the Piltdown man, found in England, still had its prominent position as the earliest ancestor of man, but Leakey strongly believed that in Africa something older could be found. Leakey, like Gibert, had a clear scientific agenda.[52] Therefore, like Gibert, Leakey wanted to achieve

45 "desestimó", "resta importancia", "ya ha avalado antes otros restos 'discutibles' ", *Ibid.*

46 "demostrarlo científicamente", *Ibid.*

47 Willis 1992, 51.

48 There are several biographies of Leakey's family used here, which always picture Louis that way. For the exact quotes see Cole 1975, 15 and Morell 1995, 139.

49 Morell 1995, 139.

50 Morell 1995, 140.

51 Bowman-Kruhm 2005, 51 and Morell 1995, 139.

52 For Leakey's claims and his early controversial claim see the chapter 'Disaster at Kanam', Morell 1995, 80–93.

international recognition and funding, and push forward a specific scientific claim. Like Gibert, Leakey wanted to have the most prominent scientists at 'his' congress and travelled to England and France and sent several letters in order to make this happen.[53] Later, he used the positive responses of these scientists to get funding from the Kenyan government.[54] During the congress, attendees could also admire Leakey's fossil discoveries, which were "carefully arranged".[55] Leakey and his team organised a 'safari' with more than 60 attendees to explore Olduvai Gorge, among many other sites.[56] According to Delta Willis, author of a book on Leakey's family, "the meeting itself offered a place to exchange ideas, but perhaps the greatest benefit was future research [...] Leakey's plan of interesting a wider group of scientists proved to be a profound success."[57]

Leakey's Pan-African Congress is just an example of how the organisation of a large conference can have several goals beyond gathering scientists and stimulating professional debate. Reinforcing specific and general scientific claims, boosting the significance of research projects, and recovering individual or general prestige could be some of the 'hidden' objectives of these meetings.[58] Moreover, during the conferences, organisers use several 'tools' to achieve these aims. These 'tools' are deployed in both the scientific and the public arena, creating a mixture of communication channels between both, which conferences of this kind again make quite apparent. On the following sections, 1995 Orce Conference will be explored in great detail in order to show how it was much more than a scientific conference but also a scientific public conference and how it is a great example to show these objectives and 'tools'.

∴

In 1993, Gibert began to think about holding a scientific conference in Orce in 1995. According to him, all of the work done since 1987, 'the great recovery' (the two monographs, the papers published in international journals, the new discoveries ...) justified and required the organisation of a major conference. Following Gibert's discourse, this conference would have the main aim of discussing the arrival of hominids to Europe in order to put the Orce discovery in the general context. Yet, it was also clearly intended to provide fertile ground for

53 Willis 1992, 51 and Morell 1995, 139.
54 Morell 1995, 140.
55 Willis 1992, 52 and Morell 1995, 142.
56 Willis 1992, 52 and Morell 1995, 142–143.
57 Willis 1992, 53.
58 Carandell 2015a and Carandell 2016.

getting his position regarding the hominid presence in Orce finally accepted by the international scientific community.[59] At that time, being at odds with the other *Institut* researchers, Gibert had little support from the Spanish scientific community. In order to get support, he turned to the international scientists that he had met in conferences around the world. He sent several letters to prominent scientists in order to invite them to be part of the conference's Scientific Committee. As Gibert explained to García-Olivares, the reply from these scientists was "very positive" and they were "looking forward to getting to know the sites and accepting the proposed paradigm."[60] This last term was often used by Gibert in the very Kuhnian sense of a revolution in the discipline. He believed that his claims of a hominid presence in southern Spain approximately 1.5 million years ago and their arrival via the Strait of Gibraltar crossing was a new 'paradigm' in studies of the hominid colonisation of Europe.[61]

With regard to the international scientists, Gibert invited, for instance, Richard Leakey, who he had probably met at the Southampton conference in 1986 and maybe again in his travels to Kenya.[62] Unfortunately for Gibert, in June 1993, just at the time when he was trying to contact him, a plane piloted by Leakey crashed and he was badly injured, losing his legs below the knee. Gibert also tried Meave Leakey, Richard's wife, but it seems that the accident prevented both from accepting the invitation.[63] Gibert tried to aim high in his ambitions regarding those scientists he believed should form part of the Scientific Committee. Gibert did not want to hold a standard conference. He was preparing something else, and for this, he needed internationally renowned scientists that could not only convince other colleagues at home and abroad, but also attract the media and politicians to the conference. Gibert needed science stars, and at that time, in palaeoanthropology, at least from Gibert's perspective, Leakey was the greatest of them all.

After Leakey's refusal, Gibert achieved two top palaeoanthropologists' acceptance as invited guests: the French Yves Coppens (figure 11) and the South African Phillip Tobias (figure 13), who had met Gibert at the 1989 conference in Turin.[64] In Spain, Gibert managed to get Emiliano Aguirre, former director

59 AJG-ICPMC: Letter from Gibert to García-Olivares, 1 October 1993 and Gibert 2004, 88–94.

60 "muy positiva", "con ganas de conocer los yacimientos y aceptando el paradigma propuesto", AJG-ICPMC: Letter from Gibert to García-Olivares, 1 October 1993.

61 See for instance: J. A., *Diari de Sabadell* 1986c and Gibert 2004, 56.

62 AJG-ICPMC: Letter from Gibert to Meave Leakey, 3 December 1993.

63 *Ibid.*

64 Gibert 2004, 78. More on Coppens: Anonymous 2009 and on Tobias: Tobias 2005 and Goodrum 2013.

of the Atapuerca project, and José Maria Bermúdez de Castro, one of his disciples and then co-director of the Atapuerca project, to attend the conference. Agustí and Moyà-Solà did not attend. Henry de Lumley, although he corresponded cordially with Gibert and even seemed to consider attending, in the end did not.[65] Eudald Carboncll and Juan Luis Arsuaga, the other scientists in charge of Atapuerca, did not attend either, despite being invited by Gibert. The comparison of the Andalusian research with Atapuerca became very popular during the Orce conference. Later, the conference attendees and their significance will be analysed in greater depth.

The conference was also announced in the media. In July 1994, for instance, *ABC* published an article entitled 'The Orce Man Resurrection'. In this article, the journalist explained the lack of excavation permits Gibert's team had and the recent international publications, finally highlighting that a great international conference, with 35 of the most prestigious scientists in palaeoanthropology in the world, would be organised the following year.[66] In some of this early accounts of the organization of a conference, it is presented as the recovery of the conference that was cancelled in 1984 when the Orce controversy exploited for the first time in the *El País* newspaper.[67] The Orce Man was alive and well again after Agustí and Moyà-Solà's attack in 1987. More than seven years were needed to bring it back to the press headlines. The conference seemed to be the final result of the Orce Man's 'resurrection'.[68] Michael Walker's discoveries in Cabezo Gordo also received a considerable amount of media attention that same year, especially in Andalusia and Murcia.[69] It was compared with Atapuerca as a key human origins site in Spain. With Cabezo Gordo in the media, Gibert did not miss the opportunity to promote Orce and Cueva Victoria as the 'real' First European sites.[70]

Yet, looking at private correspondence between Gibert and his collaborators of the period before the conference, we can find a brief letter signed by both Palmqvist and Martínez-Navarro (figure 12) that presented to their boss serious doubts about the validity of the claims made around VM-0, the Orce Man. These collaborators who had and will be co-authors of papers devoted to the

65 AJG-ICPMC: Letter from de Lumley to Gibert, 20 October 1994; Letter from de Lumley to Gibert, 24 April 1995; Letter from de Lumley to Gibert, 2 May 1995; Letter from Gibert to de Lumley, 29 June 1995.

66 'La "resurrección" del "hombre de Orce" ', De León-Sotelo, *ABC* 1994.

67 'La "resurrección" del "hombre de Orce" ', De León-Sotelo, *ABC* 1994.

68 The TV Documentary *Pedres que parlen* also presented the conference as Gibert and the Orce Man's revival: Guàrdia/Pou 1996.

69 For instance: Hurtado, *ABC* 1994 and S.C., *ABC* 1994.

70 S.C., *ABC* 1994.

FIGURE 12 Paul Palmqvist Barrena in 2014, left and Bienvenido Martínez-Navarro in 2019, right
PHOTOS PEPEEFE

Orce Man said to Gibert that the evidence was weak and that it was better to be stress the palaeontological value of the Venta Micena, and Orce generally, sites than keep on only highlighting the relevance of the Orce famous remain.[71] Palmqvist and Martínez-Navarro also complained about the lack of funding for their projects and Gibert's for them excessive leadership. These apparent divisions within Gibert's team will not affect the Conference but will have tragic consequences after that.

Just four months before the conference, in May 1995, Gibert gave a talk in the *Museu d'Arqueologia de Catalunya* in Barcelona. He started the talk by reviewing the history of the controversy. In a very typical 'heroic narrative', Gibert continued stating that despite the "great shortage of resources", his team had worked hard achieving many publications in international journals and several very promising and revolutionary studies on the Orce bone.[72] Gibert also presented the conference, and its authority, as a validation for his claims, and pointed out that large events like this conference were very useful for the development of the towns that host them.[73] He also cited other controversies

71 AJG-ICPMC: Letter from Palmqvist and Bienvenido Martínez-Navarro to Gibert, 16 May 1994.

72 "gran falta de recursos". AJG-ICPMC: Presentation of the 1995 Conference in the Museu d'Arqueologia de Catalunya, 30 May 1995. An example of the use of this kind of heroic narrative in science could be Hevly 1996.

73 *Ibid.*

in the history of palaeoanthropology that for him illustrated how common such controversies are when discoveries challenge established theories. He talked about the rock paintings discovered in Altamira in Spain, which according to Gibert were "called into question by all of French prehistory, which represented orthodoxy at that time."[74] Another controversy cited was the Taung Boy discovery by Raymond Dart in South Africa.[75] Gibert used these historical examples several times to justify the reasons for the outbreak of the Orce controversy. Like these examples, the Orce Man discovery changed scientific 'paradigms' and was actively denied by the 'establishment'.

According to Gibert, one of the main aims of the conference was to generate a dialogue with those who were sceptical about his claims, or even with those who were completely against them. Moreover, Gibert considered that the conference would be very useful for discussing hominid arrival to Europe in general.[76] Yet, while Gibert's discourse is full of 'dialogue' and 'scientific discussion', the conference documents clearly show how his main aim was not to 'debate' but to 'convince' his colleagues. As he states in the conference proceedings, "the important thing was to strengthen the Orce paradigm."[77] Similarly, Camilo José Cela-Conde, human evolution professor at the *Universitat de les Illes Balears* and conference participant, noted, "the obvious but unstated purpose of the conference was to conduct a general discussion about the Orce findings, in order to gain recognition from the international palaeontology community."[78]

On 1 September 1995, just two days before the conference, Gibert received a fax with very bad news. Yves Coppens, one of the leading scientists expected at the conference, would finally not attend due to family issues.[79] This was a heavy blow to Gibert's ambition. A great star was lost at the very last minute, and it was not the only main character to be missing. Around the beginning of

---

74 "puestas en duda por toda la prehistoria francesa, depositaria, en aquel momento de la ortodoxia", *Ibid.* See Introduction. For a critical look at the Altamira case see Moro Abadía/Pelayo 2010.

75 *Ibid.* See Introduction. More details on Dart in Richmond 2009.

76 These statements appeared in several letters to the Conference guests, for instance AJG-ICPMC: Letter from Gibert to Gerhard Bosinski, 30 September 1993; Letter from Gibert to Ofer Bar-Yosef, 3 November 1993; Letter from Gibert to Jerold Lowenstein, 2 December 1993.

77 "lo importante era consolidar el paradigma", Gibert 1999a, 10. The Conference Proceedings: Gibert et al. (eds.) 1999b.

78 Cela-Conde was also son of the Nobel Prize in Literature laureate, Camilo José Cela. "El propósito no declarado pero obvio de la conferencia era el de llevar a cabo un debate general acerca de los hallazgos de Orce, con el fin de obtener un reconocimiento de la comunidad paleontológica internacional." Cela-Conde 1996, 33.

79 AJG-ICPMC: Fax from Coppens to Florentina Sánchez and Josep Gibert, 1 September 1995.

1995, the Queen of Spain, Sofia, accepted the role of honorary chair of the Orce conference. Having studied her degree in archaeology in Greece, the Queen had always shown interest in the sciences, so the Orce conference was another initiative in which she was invited.[80] However, in the end she did not attend the conference due to other official commitments. Despite these last minute losses, the 1995 Orce International Conference in Human Palaeontology was ready to begin. Before continuing with the conference itself, let us try to understand the political relations that lay behind it.

∴

For the summer of 1995, Gibert again asked for permission to excavate in Venta Micena. He used the organisation of the conference as his main argument for receiving the permit.[81] He finally got the permission that he had not received for the last five years, since 1990. Besides Venta Micena, Gibert's team excavated in yet another site in the Orce Basin (Barranco León), where stone tools and a supposed hominid tooth were found. Nevertheless, getting the excavation permit after five years was not Gibert's only achievement. In order to organise the conference, he managed to get over 5 million pesetas (more than 30,000 euros) from the *Junta de Andalucía* and 2 million from the *Diputación de Granada*, plus one and a half million from the central government science ministry.[82] This money allowed Gibert to invite several international scientists to Orce. The proceedings were published some years later thanks to money provided by the *Diputació de Barcelona*.[83] However, why did Gibert have so many difficulties in getting excavation permits and funding before 1995, and then later got those permits and that funding from the *Junta*? The answer to this question is in many ways political and originates back in the early days of the controversy.

After the first controversy, around 1986, the two main scientists from Granada that were working with Gibert (Isidro Toro and Pascual Rivas) left the team. From then on, they became very critical of Gibert's management. Both were members of the *Partido Socialista Obrero Español* (PSOE), the socialist party that governed Andalusia during those years. In 1984, Toro became the archaeologist for the *Junta* in Granada's provincial Department of Culture. Rivas,

80 OCVR 1995 and Gibert 1999a, 10.

81 AJG-ICPMC: Letter from Gibert to Manuel García León, 2 May 1995.

82 AJG-ICPMC: Conference expenses sheet, no date, from now on n.d.

83 ADB: Letter from Gibert to Manuel Royes, 8 February 1993. The Conference Proceedings: Gibert et al. (eds.) 1999b.

besides being rector of the *Universidad de Granada* between 1989 and 1992, became a member of the Archaeology and Heritage Committee for Andalusia between 1985 and 1996. According to *ABC*, in 1994 Rivas was in charge of this committee, which wrote the negative reports that led to the Orce excavation permit denial.[84] Naturally, neither Pascual Rivas nor Isidro Toro attended the conference. According to Lluís Gibert, since 1986, and with the exception of 1990, these two characters and their political influences blocked all of Gibert's requests for excavation permits in Venta Micena and for funding from the *Junta de Andalucía*.[85]

But something happened in the Andalusian elections in 1994: the PSOE lost their absolute majority. Although its candidate, Manuel Chaves, became the president, it was a very weak government. From 1994 until 1996, the second political force, the *Partido Popular* (PP), the Spanish conservative party, and the third political force, *Izquierda Unida* (IU), a left-wing coalition set up in 1986, unofficially joined forces in what was known as the 'clothes peg' (*pinza*).[86] During those years, this strange combination of power between the conservatives and the left blocked several PSOE proposals and put forward several others against the governing party's wishes. One of the proposals moved forward without the government's approval was to give support to the Orce conference. Several reasons led IU and the PP to support the Orce research project. Firstly, Josep Gibert was a member of the Catalan *Partit Socialista Unificat de Catalunya* (PSUC) and later of the communist *Partit dels Comunistes de Catalunya* (PCC). Both political parties later joined *Iniciativa per Catalunya* (IC), the Catalan version of IU.[87] Secondly, the conservative party's desire to go against the PSOE also led to its support. The PP's support for Gibert would continue after the 1995 conference. Thirdly, Gibert's own strategy was fundamental.

Sometime before the conference, Gibert sent a letter to Francisco Ríos, IU representative in the *Junta de Andalucía*. In this letter, Gibert presented a legislative proposal to Ríos with the title 'The Development of Orce and its Surroundings through a Prehistoric Park'.[88] Gibert first explained the uniqueness of Orce and his scientific team's achievements and then presented other

---

84 Ruiz Antón, *ABC* 1994.

85 Interview with Lluís Gibert 2012. See also: Ribot, *El Ideal de Granada* 2007.

86 Durán 2012.

87 Wikipedia Contributors 'Partit Socialista Unificat de Catalunya' and Wikipedia Contributors 'Partit dels i les Comunistes de Catalunya'.

88 'Desarrollo de Orce y su entorno a través de un parque prehistórico' AJG-ICPMC: Letter from Gibert to Francisco Ríos, n.d.

examples of prehistoric parks in Europe, mainly in Tautavel in France, where de Lumley worked. Finally, Gibert requested the creation of a committee formed by members of the *Junta*'s Department of Culture, the *Diputación de Granada*, the Orce Town Council, and the scientific team to study the further development of Orce through its prehistoric heritage. The committee, concluded Gibert, could take advantage of the presence of "prestigious foreign scientists" in Orce during the conference to obtain their views on the Orce project.[89] According to *ABC*, Francisco Ríos championed the Orce cause in the parliament, where he complained about the refusal to grant permits to Gibert, asked for funding for the conference, and achieved the *Junta*'s commitment to spend 500 million pesetas (around 3 million euros) on building a palaeontological park in Orce together with a research centre and a museum. This money was never invested and the project was never implemented.[90] When Gibert finally got the funding from the *Junta* to organise the conference, he sent a fax to Francisco Ríos thanking him for everything he had done and informing him that he remained at his disposal. Gibert also stated, "sincerely, I think that the unblocking [of funding] is due to your interventions in the parliament."[91] Gibert, but also Bienvenido Martínez-Navarro and the Orce mayor, Leandro Castellar, send letters to the *Diputació de Barcelona* president those days, Manuel Royes, in order to ask him to support the Conference, to fund the publication of the proceedings, and therefore support the development of the little town of Orce and the scientific research performed there.[92]

As Agustí and Lluís Gibert acknowledged, the 'clothes peg' formed by IU and PP against PSOE allowed the Orce conference to go ahead. This entirely political factor provided Gibert with the possibility of getting resources to develop his research strategies and to get the Orce Man recognised once again.[93] Therefore, while Gibert claimed that scientific publications and other achievements provided positive conditions for organising an international conference, it was the favourable political conditions that finally allowed funding and permits for the conference to take place.[94] As Granada's provincial governor stated during

89 "Prestigiosos científicos extranjeros", *Ibid.*

90 Ruiz Antón, *ABC* 1994.

91 "Sinceramente creo que el desbloqueo se debe a tus intervenciones parlamentarias", AJG-ICPMC: Fax from Gibert to Paco Ríos, 16 March 1995.

92 ADB: Letter from Gibert to Manuel Royes, 8 February 1993; ADB: Letter from Bienvenido Martínez-Navarro to Manuel Royes, 10 December, 1992; Letter from Leandro Castellar to Manuel Royes, 21 January 1993.

93 Interviews with Agustí 2012 and Interview with Lluís Gibert 2012.

94 Gibert 2004, 88.

the opening session of the Orce conference: "the political aspect [of the controversy] is inseparable from the scientific project".[95]

3.2 An International Conference as a 'Tool' to Convince

The Orce International Human Palaeontology Conference took place over five days from 3 to 7 September 1995. The conference had a very tight and 'Spanish' timetable, holding sessions approximately from 9 a.m. to 9 p.m., with a two-hour break for lunch. The attendees also went on several excursions: on day one to visit the Orce Basin; on the morning of day three to visit several sites near Orce, including Venta Micena; and a final post-conference excursion to the Murcia region to visit Cueva Victoria and Cabezo Gordo (see map 1).

The scientific programme contained four thematic sessions and four round tables. The thematic sessions were devoted to 'Palaeontology, Taphonomy, and Anthropic Action', 'Human Palaeontology', 'Human Occupations in the Middle East, Java, Georgia, and Europe', and 'Human Remains, Anthropic Action, Fauna, and Geology in Orce, Cueva Victoria, and Cabezo Gordo'. Gibert and his team reserved a complete thematic session divided into the first afternoon and the morning of the second day for their discoveries, one fourth of the presentations were thus about the occupation of southern Spain. The round tables dealt with 'Hominid Dispersion', 'Fauna Migrations', 'Hominids of Eurasia', and 'Human Occupations of Europe'. Tobias, together with Campillo, assumed the responsibility of holding a plenary session at the beginning of the event.[96]

Around 200 scientists attended the conference. Next section was made using Gibert's own final list of conference attendees.[97] At that time, Orce had around 1500 inhabitants, so almost the equivalent of a seventh of the population came to the town for the conference, to which we must also add journalists, their assistants, politicians ...[98] Of those 200 scientists, 24 were from Gibert's and Walker's team. Around half of the attendees (105) were Spanish, 72 of which were students. France, with 16 attendees, and Italy, with 15, were the countries best represented, but there were also scientists from the United Kingdom, the United States of America, Germany, Israel, South Africa,

95 "el aspecto político es inseparable del proyecto científico", OCVR 1995.
96 AJG-ICPMC: 1995 Orce International Human Palaeontology Conference Programme, n.d.
97 AJG-ICPMC: List of conference attendees, 1 March 1995.
98 Wikipedia Contributors 'Orce'.

Kenya ...[99] The conference programme was written in Spanish, French, and English and all the talks had simultaneous translation from Spanish to English or from English to Spanish.[100] The funding for the conference allowed the organising committee to invite the main scientists. Around 23 of the attendees were members of the conference's Scientific Committee and therefore had their travel and accommodation paid for.[101] In addition, there were also around 20 invited scientists with their expenses covered. In total, adding the local committee and some student grants, around 60 scientists attended the conference with all expenses covered. Therefore, some 140 scientists paid for their trip to Orce, their accommodation, and their fees.

Some invited scientists finally did not attended to the Conference. They included the Leakeys, Yves Coppens, the English scientists Leslie C. Aiello and Chris Stringer, Henry and Marie-Antoinette de Lumley, and Jordi Agustí.[102] The official scientific committee of the Conference included, aside from Gibert's close collaborators, Tobias and Aguirre, scientists like Derek Roe, British expert in stone tools from *Oxford University*, who collaborated with Gibert and actively corresponded with him; Clark Howell, an American leading palaeoanthropologist with several pioneering excavations in Africa and Spain; Ofer Bar-Yosef, archaeologist from Harvard University, who performed famous research on several Neanderthal sites in Israel; Eugène Bonifay, French advocate of an over 2 million-year-old hominid presence in Europe; Vera Eissenmann, who studied prehistoric horses in Africa with the Leakeys and those of Venta Micena, supported Gibert's claims and actively corresponded with him; Jaques Tixier, French prehistorian that published the Fuente Nueva 3 stone tool findings with Gibert's team; Gerhard Bosinski, member of the team that described the first hominids from Dmanisi, who actively corresponded with Gibert and Assumpció Malgosa, Spanish anthropologist who participated in the anatomical study of some of the supposed Orce hominid remains.

The list of the attendants tells us many other things. Firstly, despite last minute casualties, leading international and national scientists did indeed gather at the Orce conference. Tobias, Howell, Aguirre, the *American Museum of Natural History* Palaeoanthropologist well known for his popular science books Ian Tattersall or Milford Wolpoff, well-known palaeoanthropologist and main proponent of the multiregional model, were among the

99 AJG-ICPMC: List of conference attendees, 1 March 1995.

100 AJG-ICPMC: 1995 Orce International Human Palaeontology Conference Programme, n.d. Also OCVR 1995.

101 AJG-ICPMC: List of conference attendees, 1 March 1995.

102 *Ibid.*

most important scientists in human origins research in the 1990s.[103] Secondly, the Orce conference also enjoyed the attendance of a great deal of the researchers interested in the 'First Europeans'. For instance, new findings from Atapuerca and Dmanisi in Georgia were presented at the event. Bonifay, Aldo Serge, an Italian researcher who described the Ceprano Man; Karel Valoch, advocate of a 1.5 million-year-old hominid presence in Europe and leading proponents of hominid migrations, like Wolpoff, Wil Roebroeks, proponent of the 'short chronology' of a later arrival of hominids to Europe or Paul Sondaar, who performed research on hominid arrival to Java, presented papers in Orce. The conference was a major event concerning the study of the early occupation of Europe.[104]

Thirdly, Gibert invited several scientists that were opposed in one way or another to his research conclusions, though none of them was on the Scientific Committee. Noteworthy examples include not only Agustí and the de Lumleys, who did not attend the conference, but also Wil Roebroeks and José María Bermúdez de Castro, who did. By inviting his opponents, it seemed Gibert was not avoiding scientific or public debate. The conference had to be widely followed by Spanish media, so, even if this could turn against him, Gibert openly accepted differences of opinion within the conference.

In the press, Gibert even described Agustí's absence as "unacceptable" and "regrettable".[105] Agustí was the then director of the *Institut de Paleontologia de Sabadell*, where Gibert worked, and he also conducted biostratigraphic research (his speciality) in the Orce area. Agustí alleged that he had to attend another conference in Romania at the same time.[106] In 1984, Marie-Antoinette de Lumley's statements in *El País* that the Orce Man was an *Equus* forced Gibert to go public and to defend himself and his research. Now Gibert, with the backing of an international conference, was forcing his opponents, especially Agustí and the Atapuerca researchers, to appear in public and give their opinion on his claims. Moreover, forcing them in the very favourable environment for him that the conference was. This strategy led first to positive statements from supposed 'opponents' like Bermúdez de Castro. However, it also led to negative statements and a later 'attack' against Gibert. In some ways, he was exposing his credibility to the public. As Gibert saw it, by accepting that the

103 For more on Howell see: Klein 2013; On Tattersall: Anonymous 'Ian Tattersall'; On Wolpoff: Wikipedia Contributors 'Milford H. Wolpoff'. According to Nora Moloney, an English PhD student who attended the Conference, it "was well attended with many well known palaeoanthropologists and archaeologists". Interview with Nora Moloney 2014.

104 Interview with Lluís Gibert 2012.

105 "inadmisible", "lamentable", Ache, *Diari Sabadell* 1995.

106 Ache, *Diari Sabadell* 1995 and Interview with Agustí 2012.

Orce Man was in fact a hominid and that Orce held traces of the first inhabitants of Europe, the scientific community was also accepting him as a valid and credible 'expert'.[107]

∴

By forcing his opponents to appear in public, Gibert was trying to achieve his 'hidden' aim: to get public recognition, and therefore political recognition, and therefore permits and funding for his research. If opponents and international scientists publicly recognised the significance of the Orce research project, then the general public would finally realise that its significance was real. Crucially, this public also included politicians. The Orce research was again presented as scientifically very important but, at the same time, as having serious problems regarding funding and excavation permits. How did Gibert manage to 'convince' all of these 'publics' that he was addressing? To analyse Gibert's strategies during the conference, I will use the terms 'tools' or 'resources', following sociologist Brian Martin's use in his analysis of the fluoridation controversy.[108] Other sociologists of science such as Harry Collins, Bruno Latour, Stephen Woolgar, Michael Mulkay, and Nigel Gilbert have used these or other terms in a similar way in their work on scientific controversies.[109] Martin defined 'tool' or 'resource' as "anything that is used by an actor, meaning, in this instance, someone or some group involved in the controversy."[110] For Martin, in controversies, it is more useful to think of evidence and arguments as 'tools' than to think of them as "statements of reality".[111] Beyond evidence, these resources could also include "scientific publications, scientific status' and '[...], professional prestige, authoritative endorsements, community organizations, governments, and the mass media."[112] These 'tools' are used then in scholarly rhetoric to convince colleagues of certain controversial scientific claims. Both in public and scientific fields, 'tools' are also used to "recruit supporters and denounce opponents."[113] Gibert, and other actors in this story, used these 'tools' from the very beginning and throughout the development of the controversy. Yet, the 1995 scientific gathering, with its intense media coverage

107 See a similar example with Dart and the Australopithecus in Richmond 2009, 11.
108 Martin 1991, 8 and 156.
109 Collins 1981; Knorr/Krohn/Whitley 1981, 139–163; Latour/Woolgar 1979; Mulkay/Gilbert, 1981.
110 Martin 1991, 110.
111 Martin 2014, 156 and Martin 1991, 110.
112 Martin 1991, 5 and 110.
113 Martin 2014, 34.

and our access to internal documents from the organisation of the event (including video recordings), is of special interest for this 'resource' analysis. In this short period of time, just some days in the beginning of September, the 'tools' and strategies are more visible for historical analysis than in a longer controversy. These resources could be verbal or visual, but also material or even living entities and could be directed at and used by different actors in the controversy.

The Orce conference began on 3 September 1995 with an excursion to the Baza area. As mentioned earlier, the conference featured three excursions visiting sites like Venta Micena, Cueva Victoria, Fuente Nueva 3, and Barranco León. Site visits are common in archaeological and palaeontological conferences. However, following Martin's analysis, can be interpreted that the organising committee's desire to show all of the Orce sites and to explain, *in situ*, their geological, palaeontological and archaeological characteristics not only as 'showing reality' but as one more attempt to present their claims to the conference attendees in order to achieve their goals. As Gibert stated in his book, the site visits "ensure[d]" the success of the conference, which aimed to "convince" the participants of the Orce "paradigm".[114] The visits also featured explanations by Gibert's team, such as, for instance, Bienvenido Martínez-Navarro at Fuente Nueva or Lluís Gibert at Barranco León.[115] As the archaeology historian, Margarita Díaz-Andreu notes in her book on the relationship between Spanish and English archaeologists, "the personal experience of being at a site was a very effective means of advertising its significance."[116] In his popular science book, Gibert quoted scientists such as Derek Roe or Gerhard Bosinski who after one of these visits to Orce 'reconverted' to his side.[117] Journalists and politicians, like José María Martín Delgado, head of the *Junta*'s Department of Culture or Diego Valderas, president of the Andalusian parliament, also visited the Orce area sites with members of Gibert's team.[118] Televisions also attended the excursions, showed the visits and provided explanations on the news.[119] In short, the site visits of the Orce conference 'show' were one of the very useful and persuasive 'tools' used to get scientific and also public supporters for the Orce cause.

114 "asegurar", Gibert 2004, 87.
115 Gibert 2004, 168.
116 Díaz-Andreu 2012, 377.
117 Gibert 2004, 87–88.
118 *Ibid.*
119 For instance: Canal Sur 1995a; Canal Sur 1995b.

FIGURE 13 Philip Tobias in 2007
PHOTO RAASGAT

Throughout the controversy, Gibert developed a discourse of 'me against the world'. He presented his ideas about the Orce Man as a new 'paradigm' in Kuhnian terms and pictured himself as marginalised 'revolutionary'. The Orce conference showed the paradox of continuing with this discourse while at the same time claiming that Orce was internationally recognised and somehow mainstream. Gibert needed both sides of the discourse in order to get attention in both scientific and public forums. In the closing session, Gibert stated that the research project in Orce was one of the "most important human palaeontology projects in Europe".[120] Gibert also presented the conference as a turning point regarding the scientific community's opinions on his research. Everything seemed to have changed and now the broad scientific community, according to Gibert, accepted the Orce claims.[121] The isolated hero discourse and the worldwide acceptance discourse were also 'tools' that Gibert used before, during, and after the conference.

120 "uno de los proyectos de paleontología humana más importantes de Europa", OCVR 1995.
121 *Ibid.*

In his popular science book, Gibert admitted that the way he structured the conference talks was a "scientific strategy".[122] Again, it is useful to see scientific talks as much more than the mere presentation of scientific knowledge. With the goal of convincing the attendees, Gibert put all of the Orce papers together in the afternoon of day one and the morning of day two.[123] These presentations filled more than a quarter of the whole conference; 18 out of approximately 68 talks.[124] This 'scientific strategy' began in the second plenary session with Domènec Campillo lecturing on the human endocranium "with clear allusions to the Orce cranium", according to Gibert.[125] After lunch, Gibert dealt straight off with the Orce hominid remains; Palmqvist presented the recently published fractal study; García-Olivares and Jerold Lowenstein showed the immunological results; Lluís Gibert, then a young postgraduate student, explained the geological characteristics of Barranco León and its stone tools; and Alain Turq presented Fuente Nueva 3 and its lithic industries. The session the following morning was devoted to Orce fauna, with Palmqvist and Bienvenido Martínez, among others, and to the Cabezo Gordo hominid remains, mainly presented by Michael Walker. According to Gibert, "the strategy worked well".[126] To sum up, the organisation of the conference talks was used by Gibert and his team as another 'tool' in their strategy to win followers for the Orce 'paradigm'.

During the conference, the Orce hominid remains and the stone tools were put on display near the venue where not only the invited scientists but also journalists and the general public could see them. According to Gibert, "these valuable remains were closely examined by all the experts" and a "private session with the best palaeoanthropologists" was scheduled to discuss them.[127] During that session, Gibert experienced an "exciting moment" when Tobias (figure 13) examined the piece and discussed it with Campillo.[128] For him: "If Tobias accepted its human nature, the conference would be a success. And that's how it was".[129] This private moment also had its public equivalent. Televisions and newspapers showed the remains to readers and audiences. In an interview for the Spanish television channel Canal Plus, Phillip Tobias talked

122 "estrategia científica", Gibert 2004, 93.

123 AJG-ICPMC: 1995 Orce International Human Palaeontology Conference Programme, n.d.

124 *Ibid.*

125 "con claras alusiones al cráneo de Orce", Gibert 2004, 93.

126 "La estrategia resultó bien", *Ibid.*, 93.

127 "Estas valiosas piezas fueron observadas con atención por todos los especialistas", "sesión privada con los mejores paleoantropólogos", Gibert 2004, 89.

128 "momento emocionante", "la pieza", *Ibid.*, 89.

129 "Si Tobias aceptaba su naturaleza humana, el congreso era un éxito. Así fue", *Ibid.*, 89.

to the journalist in front of the Orce remains and even showed them to the audience.[130] Cela-Conde stated in a review of the conference for the Spanish version of *Scientific American* that during the attendees' visit to the Barranco León site, Gibert's team presented BL-O (a supposed hominid tooth fragment that had not been yet published) to the scientists as final evidence of the validity of their claims.[131]

All of these examples lead us to the concept, introduced by the historians of science David Van Reybrouck, Raf de Bont, and Jan Rock, of 'material rhetoric'. For these scholars, material rhetoric is a "regime of material practices that scholars deploy to accompany traditional, verbal, and pictorial rhetoric with which they distribute their theories and ideas." For them, among other practices, "displaying finds" is one of the "tricks and tropes of this non-discursive, material form of rhetoric".[132] The academic authority given to Gibert and his team due to their control over these research objects allowed them to display the scientific material as what they claimed it to be. The hominid remains were displayed as hominid remains, with a prominent location for VM-O: the famous Orce Man. Therefore, during the conference, the display of these controversial remains and artefacts was an important part of Gibert's rhetoric, a crucial 'tool', to convince colleagues and the public of his claims.[133]

The prestigious scientists that attended were also used as part of Gibert's tactics to convince. He stated in the opening session that all those who had accepted to be part of the conference's Scientific Committee were also, with that acceptance, "backing our research".[134] This statement shows how Gibert understood the international scientists' presence at Orce as support for his controversial claims. The Queen of Spain's acceptance of the role of honorary chair of the conference was also used as a great source of prestige for the event.[135]

To sum up, a wide range of these 'tools' used during the conference to disseminate and establish Gibert's main claim that hominids were living in Orce at least one and a half million years ago and that the controversial cranial fragment VM-O was proof of this. In a way, the whole conference was a huge 'show', a 'performance', where these 'tools', present throughout the whole controversy,

130 Canal Plus 1995.
131 Cela-Conde 1996.
132 Van Reybrouck/de Bont/Rock 2009, 198.
133 We can find another good example of the use of material rhetoric in Richmond's reconstruction of Raymond Dart's Australopithecus findings. Richmond 2009, 67–81.
134 "Apoyando nuestra investigación", OCVR 1995.
135 OCVR 1995 and Gibert 1999a, 10.

were used in a more concentrated and visible way. This 'performance' had the main aim of gaining authority in scientific practice and knowledge production and, thus, 'constructing' the reality of Gibert's claims.[136] Somehow, the whole Orce conference '(re)constructed' the Orce Man reality, destroyed after years of public controversy. Even more importantly, this reconstruction was made in a scientific forum, but in a very public and open way. In addition, for Gibert, the conference somehow represented the recovery of his research and claims, but also of his own honour and credibility as a scientist. Its success also meant validity and success for his research project, his practices and methods and his scientific claims.

During the Orce conference, a number of screens were put in Orce's main square to allow those that did not fit in the conference venue (mainly Orce's inhabitants) to follow the event.[137] The palaeontological remains were also displayed for the people of Orce to see. During the years after the controversy erupted in 1984 and despite his dispute with Spanish academics, Gibert established a close relationship with Orce, its politicians, and its inhabitants. Often, he tried to gain the locals' favour by highlighting the huge scientific and, above all, economic potential of his excavations in Orce. Gibert proposed the construction of a prehistoric and palaeontological park in Orce in order to boost tourism.[138] Gibert spent every summer in Orce since the early 1980s and was nominated 'adopted son' of the town in 1986. A little later, a petition to give Gibert's name to the local museum was accepted after signatures were collected from Orce's inhabitants.[139] As the Argentinian historian of science Irina Podgorny suggested in the South American context, European researchers defined themselves as natives in order to legitimate themselves as state officials. This legitimisation also allowed the circulation of their claims among colleagues and the public.[140] Another good example could be the German archaeologist Manfred Korfmann, who worked in Hisarlik in Turkey, the alleged location of the ancient Troy. It seems that Korfmann, who even took on the second first name 'Osman', "maintained a special relationship with Turkish academia and

136 These 'performative aspects of authority' in scientific practice are further analysed in Pruitt 2011, 200–233.

137 OCVR 1995.

138 AJG-ICPMC: Letter from Gibert to Francisco Ríos, n.d.

139 Gibert 2004, 425.

140 Podgorny 2005, 264.

politics", which apparently granted him "exclusive excavation rights" for the site.[141] Gibert, who came from Catalonia, somehow wanted to 'become a local' and even tried to get a permanent job at the *Universidad de Granada*, the nearest university to Orce. Yet, according to Gibert, Pascual Rivas, the palaeontology professor at that university, blocked his application.[142]

Moreover, from the discovery and the early controversy all the way to the screens in Orce's main square, Gibert had successfully controlled the town's public opinion, which was always favourable towards him. This can be seen in the way the mayor of Orce, Leandro Torres, referred to Gibert during the conference inauguration session: "Our beloved José Gibert".[143] In a way, with all the controversy in the media, the lack of support in the *Institut*, the lack of support from the rest of the Spanish scientific community, and the lack of support from Andalusian politicians, the town of Orce became a safe place for Gibert, a place where criticism and rejection were less likely. As he stated in the media after his nomination as Orce's adopted son, "[Orce's inhabitants] have helped me work happily".[144]

Nevertheless, this support was not easy or free, an intense popularisation effort was made by Gibert and his team within the Orce town. According to Gibert, by 2004, he had given up to 17 public talks (or "informative talks", as he called them) in this small town.[145] In 1991, Gibert and his team also promoted a trade school for young people in the town in order to renovate an old palace and use it for the museum. Students learned masonry, plumbing, or carpentry but also museology and restoration of archaeological and palaeontological remains.[146] Just after the conference, another good example of how Gibert used the town's promotion as a way to get, in this case, support from Orce's politicians was raised. In order to justify the efforts the Orce town had made for his research and during the conference, Gibert told José Ramón Martínez Olivares, Orce's deputy mayor: "I'm promoting Orce. There will be a report in *Science News*, it will appear in the *National Geographic*, the *American Journal of Physical Anthropology*, the *Journal of Human Evolution* ..."[147] So, Gibert highlighted the appearance of his research in highly specialised journals as well as popular magazines to show how it would help the development of Orce. The direct

141 Wagenknecht 2012, 295 and 300.

142 AJG-ICPMC: Letter from Gibert to Enrique García-Olivares, 1 October 1993.

143 "nuestro querido José Gibert", OCVR 1995.

144 "me han ayudado a trabajar a gusto", MacPherson, *El Periódico* 1986.

145 "Charlas informativas", Gibert 2004, 53.

146 Martínez-Navarro 1993, 49–50.

147 AJG-ICPMC: Letter from Gibert to José Ramón Martínez Olivares, 23 October 1996.

results of scientific research (publications) were therefore presented to local politicians as an achievement.

To sum up, Orce became Gibert's stronghold, allowing him to keep his Orce Man alive through several initiatives and achievements, including the Orce conference. In this periphery within the periphery, at this local level of the previously discussed geographical dimension of the controversy, Gibert became a powerful figure, a well-known character representing science (or at least palaeontology and human origins studies) for Orce's inhabitants. This was a mutualist relation. The Orce town also gained tourist attractions and funding by supporting Gibert's controversial claims. VM-0 was always present at Orce's local museum. After the Conference, even an Orce street was named after the scientist Phillip Tobias (figure 13), one of the main international supporters of Gibert's claims.[148] This was a local perception of science that contrasted, in its unconditional support for Gibert's claims, with the views from Andalusia, Sabadell, and the whole country, where everything from Orce was perceived, to say the least, as controversial.

3.3 A Triple Victory (Science, Media and Politics)

Spanish newspapers and television followed the Orce conference almost daily. During the week of the conference, the newspaper *ABC* published up to 10 articles and brief notes on Orce in its national and its Andalusian editions. *El Periódico* published five, *El País* published four, and *La Vanguardia* published three longer pieces.[149] On television, Orce was broadcast at least six times on the Andalusian channel Canal Sur.[150] The same channel also devoted a fifteen-minute report in the investigative programme *Los Reporteros* to Orce, with the conference as a main topic.[151] The nation-wide public television channel TV1

148 AJG-ICPMC: Letter from Tobias to Gibert, 18 November 1999.

149 De León-Sotelo, *ABC* 1995b; F.R.A. *ABC* 1995; Ruíz Antón, *ABC* 1995b; Troyano, *ABC* 1995; Ruíz Antón, *ABC* 1995c; S. C., *ABC* 1995; Redacción, *ABC* 1995a; Hurtado, *ABC* 1995; Ruíz Antón, *ABC* 1995a; Redacción, *ABC* 1995b; Arias, *El País* 1995a; Arias, *El País* 1995b; Arias, *El País* 1995c; Cervera, *El País* 1995; Del Hoyo, *El Periódico* 1995a; Del Hoyo, *El Periódico* 1995b; Garrido, *El Periódico* 1995; Del Hoyo/Madridejos, *El Periódico* 1995; Carbonell, *El Periódico* 1995; Turbón, *La Vanguardia* 1995; Redacción, *La Vanguardia* 1995; González, *La Vanguardia* 1995. See also: Aragay, *Avui* 1995. To compare the press following of the Conference with the wider following of the Orce controversy see Annex 2.

150 Canal Sur 1995a; Canal Sur 1995b; Canal Sur 1995c; Canal Sur 1995d; Canal Sur 1995e; Canal Sur 1995f.

151 Canal Sur 1995g.

devoted, at least, two news pieces to the conference. On the nation-wide private television channels Antena 3 and Canal Plus Orce was featured at least once in the news.[152] Some were long news pieces featuring interviews with some of the national and international scientists gathered in Orce and reports from the sites.[153] Internationally, the conference was reported at least in the German newspaper *Frankfurter Allgemeine Zeitung*, the British *The Independent*, and in the magazine *British Archaeology*.[154] There were also other magazines and more specialised journals that published reports on the conference.[155] Given all this media attention, it is not surprising that the Orce conference had sponsors that, for instance, posted their logos on the walls behind the speakers. The interest of the media on the Orce controversy was clear, yet it is also clear that Gibert actively looked for media attention when settled the press conference or send press releases to newspapers. To achieve its aims, the conference had to be a public conference, and Gibert had to be sure that the dominant discourse around it was his own discourse.

The majority of these reports, both on television and in newspapers, followed several lines of discourse. Firstly, they highlighted the significance of the conference and the 'definitive evidence' that the Orce Man was, in the end, the 'First European'. In their reports, journalists granted the authority to decide on the final acceptance of the Orce remains to the scientists gathered at the conference. For example, *La Vanguardia* featured a full-page article with a photo of Gibert and Tobias entitled 'An international conference ensures that the Orce Man is the oldest European'.[156] In their coverage of the conference, newspapers and television channels included several scientists' opinions, which were mainly positive with regard to the Orce research. At the same time, some journalists reviewed some of the conference sessions, especially the opening and the closing sessions, which were also partially broadcast by television channels.[157]

According to *El País*, in the opening session, the prestigious scientists Phillip Tobias and Clark Howell stated that they were "convinced that the Orce

152 TV1 1995a; TV1 1995b; Canal Plus 1995; Antena 3 1995. TV coverage of the conference was obtained thanks to Lluís Gibert. The news items were collected by the Orce conference's press office and this may make the compilation incomplete.

153 For instance: TV1 1995b and Canal Plus 1995.

154 Denison 1995; Keys, *The Independent* 1995; Weimer, *Frankfurter Allgemeine Zeitung* 1995.

155 For instance: Cohnen and Gómez, 1995 and Cela-Conde 1996.

156 'Un congreso internacional asegura que el hombre de Orce es el europeo más antiguo', Redacción, *La Vanguardia* 1995.

157 For instance, Canal Sur 1995a, or Moran, *El Ideal de Granada* 1995.

skull belongs to a hominid".[158] *El País* also highlighted the Dutch scientist Paul Sondaar's presentation at the conference. Sondaar lectured on the crossing made by Indonesian hominids of a 20-kilometre sea area to reach the Flores Island 0.7 million years ago.[159] According to the journalist, Sondaar's claims supported Gibert's own hypothesis about the earlier Strait of Gibraltar crossing.[160] In general, the media concluded that the Orce Man was not only a man but also the 'First European', using the scientists' opinions as their main evidence. The 'icing on the cake' was an article published in *La Vanguardia* ten days after the conference by Daniel Turbón, human evolution professor at the *Universitat de Barcelona* who attended the Conference, entitled 'The Orce Man: Josep Gibert was right'.[161]

It is also worth noting that, the media appearance of the Orce Man at the conference was linked to a 'lack of resources and excavation permits' discourse. Again, the media highlighted how Gibert's team had work with almost no economic support from the institutions. Not a journalist, but Phillip Tobias himself, in the opening session with politicians at the same table, found the exact quote for the media to define this situation: paraphrasing Churchill, Tobias noted how "never,' in the field of human evolution 'was so much done with so little ... money".[162] Once again, the *per aspera ad astra* strategy had worked. In the closing session, Martín Delgado (PSOE), from the *Junta*'s Department of Culture, promised resources and permits to Gibert and to the Orce research project.[163] He also noted that the committee in charge of granting excavation permits would be restructured. Diego Valderas, member of IU and president of the Andalusian Parliament, also attended the closing session of the conference. He recognised that he had been following the Orce controversy and conference through the media; finally stating that "Gibert has won the battle over public opinion and the battle with the Andalusian Parliament," in a clear reference to his party's promotion of the Orce research in that same parliament.[164]

---

158 "Convencidos de que el cráneo de Orce pertenece a un homínido", Arias, El País 1995a.

159 *Ibid.* and OCVR 1995. More on Sondaar in Reumer/de Voss 1999.

160 Cervera, *El País* 1995.

161 'El Hombre de Orce: Josep Gibert tenía razón', Turbón, *La Vanguardia* 1995.

162 "Nunca nadie hizo tanto con tan poco ... dinero", Morán, *El Ideal de Granada* 1995. The original Churchill quote referred to the British airmen and can be seen in: Wikipedia Contributors 'Never in the field of human conflict was so much owed by so many to so few'.

163 OCVR 1995 and F.R.A. *ABC* 1995.

164 "Gibert ha ganado la batalla de la opinión pública y la de la cámara andaluza" EFE, *ABC* 1995.

Valderas's statement reveals how the Orce conference was not only organised to fight in a scientific dispute, but to win other 'battles' in different areas, with other 'publics'. The conference was also conceived as a defence of the research project addressed to journalists, politicians and the general public. It was also a way to 'show off' to Orce's inhabitants, who received lots of national and international scientists, were able to follow the sessions from the Orce town square, and could also see the displayed remains. It was real proof of how Gibert's research could bring tourism and economic activity to the region. Finally, it was a result, of a strictly political struggle within the Andalusian Parliament.

In addition, at the end of the conference, newspapers also reported on the attendees' support of the request to have Orce recognised as a *UNESCO* World Heritage Site.[165] During the final day of the conference, the organising committee circulated a statement to be signed by the attendees in support of this recognition and other claims made by Gibert's team.[166] Although most of them signed, it seems that later some of them did not completely agree with this initiative or with the way it was presented.[167]

Another line of discourse followed by the media was the emphasis on the dispute between Orce and Atapuerca over which was the home-town of the 'First European'. Journalists mostly stated that up until then Atapuerca was in the lead but that the new evidence ensured victory for Orce.[168] As a *La Vanguardia* article emphasised, "Scientists participating in the international conference on the Orce findings assure that the remains found there are a million years older than those found in Atapuerca (Burgos)."[169] To consider this statement in its context, must be recalled that just before the Orce conference, the Atapuerca team had published an article in *Science* with their 700,000-year-old early Europeans.[170] José María Bermúdez de Castro, a third of the Atapuerca research directors' triumvirate, attended the Orce conference where he presented some of the remains they had found.[171] During the conference, Bermúdez de

165 González, *La Vanguardia* 1995 and Ruiz Antón, *ABC* 1995b.

166 OCVR 1995.

167 Personal Archive Jordi Agustí: Letter from Howell to Palmqvist 30 September 1995.

168 For instance, Redacción, *La Vanguardia* 1995; Cohnen and Gómez 1995; Antena 3 1995; Del Hoyo, *El Periódico* 1995a.

169 "Científicos que participan en un congreso internacional sobre los hallazgos en Orce aseguran que los restos del homínido encontrado en la localidad granadina son un millón de años más antiguos que los localizados en Atapuerca (Burgos)." Redacción, *La Vanguardia* 1995.

170 Carbonell et al. 1995 and De León-Sotelo, *ABC* 1995a.

171 OCVR 1995.

Castro also made statements to the media, avoiding the conflict between both research projects, and, without committing himself to the validity of the Orce bone, emphasising the significance of the Orce research project more generally.[172] Meanwhile, Gibert also stated that they did not want any 'competition' with Atapuerca.[173] Yet privately, for Gibert there was some dispute. Before the Conference, Gibert sent a letter to Joan Albaiges i Riera, a Catalan chemist in charge of the Catalan government research Department. Gibert complained to Albaiges that their research projects applications to the Spanish research council were usually refused. The reason of these refusals were, according to Gibert, that reviewers were always the leaders of the Atapuerca team, or very close to them. Gibert also pointed out that when international referees were involved; his team usually got the funding. Gibert asked Albaiges his help with this situation and announced him the forthcoming Orce Conference.[174]

Despite Bermúdez de Castro and Gibert's public remarks, this private confrontation between both Spanish towns for 'first place' was made explicit by the Spanish media. Additionally, the Antena 3 television channel's news programme interviewed some Atapuerca villagers to get their opinion on their 'competitor'.[175] In the same vein, the weekly magazine *Tribuna* headlined with 'Spanish palaeontologists fight over the discovery of the First Europeans', highlighting the opinion of Juan Luis Arsuaga (another of the Atapuerca co-directors) on Orce's supposed hominid remains: "they belong to animals, not to men".[176] Arsuaga compared the Orce remains with UFOs, stating, "a valid one never appears" and adding that the Orce remains that he had been able to examine "constitute a genuine *X-Files* case". In the mid-1990s, the X-Files TV series was being very successful in Spanish television.[177] The journalists then noted that Gibert "has been taken as a crazy genius by the international scientific community."[178] In *El Periódico*, Eudald Carbonell, the third member of the Atapuerca triumvirate, urged caution when assessing the Orce remains, stating that the important thing is for discoveries "to be accepted by the international scientific community." For this, he continued, research must be published in

172 For instance, Moran, *El Ideal de Granada* 1995.

173 De León-Sotelo, *ABC* 1995a.

174 AJG-ICPMC: Letter from Gibert to Joan Albaiges, 12 January 1995.

175 Antena 3 1995.

176 "Paleontólogos españoles se disputan el hallazgo del primer europeo", "pertenecen a animales, no a hombres", Cohnen and Gómez 1995.

177 "nunca llega a aparecer uno en condiciones", "los restos de Orce que he tenido la oportunidad de ver constituyen un genuino "Expediente X", afirma Arsuaga." *Ibid.*

178 "Gibert, que ha sido tomado como un genio loco por los santones de la comunidad científica internacional", *Ibid.*

journals with a high impact factor, as the Atapuerca team had done in *Science*, *Nature*, the *Journal of Archaeological Science*, the *Journal of Human Evolution*, etc. To top it all off, Carbonell finished his opinion piece encouraging "all the teams in our country to use this strategy", in a clear message to the Orce research team.[179] A month after the conference, *El País* published statements on the Orce bone by Bermúdez de Castro and Carbonell. For Carbonell, the research carried out in Orce did not follow "the international standards of scientific rigour", and for Bermúdez de Castro, who attended the conference, "according to the fossils displayed, there is no Orce Man."[180]

During the conference, the media presented the Orce research as at least as important as the Atapuerca project, or perhaps even more important. To counteract this, the Atapuerca research team (which had successfully used the media to achieve their aims throughout their work) again performed a clear example of the expulsion version of 'boundary-work'.[181] In public, Arsuaga, Carbonell, and Bermúdez de Castro wanted to show that the only 'serious' palaeoanthropological research being conducted was in Atapuerca, and that the research in Orce was not rigorous or was even just an 'X-File'. As Gieryn states: "to exclude an impostor scientist will focus attention on the poser's failure to conform to the expected method of logical or ethical standards variously mapped out as necessary for genuine scientific practice"[182] Again, the Atapuerca researchers wanted to draw a line between themselves and Gibert. Also according to Gieryn, "boundary work is a strategic practical action" that scientists use "to secure academic respectability", among other things.[183] The Orce-Atapuerca case was thus a 'fight' over respectability in scientific research. A 'fight' for the 'niche' position as the most prestigious and well-known Spanish palaeoanthropologist, for which, after the conference, Gibert was seriously applying. Carbonell and Arsuaga's statements show their desire to exclude Gibert from this possibility by labelling him as unscientific in another step in the 'construction' of Gibert's isolation.

In that same sense, in a 1996 paper in the archaeology journal *Antiquity*, scholars Robin Dennell and Will Roebroeks stated that the Orce hominid remains were "ambiguous, to say the least".[184] These scholars went on to say that

179 "sean aceptados por la comunidad científica internacional", "a todos los equipos de nuestro país a participar en esta estrategia", Carbonell, *El Periódico* 1995.

180 "A partir de los fósiles presentados, el hombre de Orce no existe", Rivera, *El País* 1995.

181 For an exhaustive account of the Atapuerca research project's history see Hochadel 2013b.

182 Gieryn 1999, 22.

183 Gieryn 1999, 23.

184 Dennell/Roebroeks 1996, 538.

"claims that they are hominid rest on fractal geometric analysis of the cranial sutures [...], canonical discriminant analysis of the Cueva Victoria phalange [...], Fourier analysis of the cross-section of the Venta Micena humerus [...] and various immunological analyses[...] all of which are highly esoteric techniques."[185] Like the Atapuerca researchers, these statements show Dennell and Roebroeks' desire to place Gibert's claims and methods outside of what they considered science or, more precisely, palaeoanthropology, in order to downplay the significance of the Orce findings in their article. Simultaneously, this quote also shows the many different studies performed on the Orce remains in efforts to validate them as important parts of Gibert's public and scientific discourse.

Unlike the plain anatomical analyses of the different characteristics of the Orce fragment, these techniques apparently provided a more 'objective' analysis of the bone and could thus help in solving the controversy from outside palaeoanthropology itself. Yet, once again, a controversial situation reveals how even what appears as more impartial science also involves a high degree of interpretation. Contamination with present day human proteins and even failure in their capacity to recover ancient proteins were the main concerns that informed criticisms of the immunological techniques.[186]

On the other hand, the fact and the way that these two 'esoteric' techniques were applied to the Orce Man cranial fragment show how disputes among scientists can generate the search for new arguments and evidence that may lead to the development of new techniques or the application of techniques from other disciplines to palaeoanthropological research. Debate among specialists and perhaps particularly public debate thus stimulated scientific research, multiplying the number of publications and the different studies carried out on the bone.[187]

To sum up, the media was in general very supportive of Gibert's claims and often quoted international scientists to support them. It seemed that Gibert's strategies and 'tools' had worked well as he managed to convince the three intended 'publics'. As Valderas noted, he won both the public and the political battles. Gibert apparently also won the national and international scientists' recognition, if not of the Orce Man bone, definitely of the Orce research project as a whole. The Atapuerca research team was almost the only publicly

185 *Ibid.*, 538.

186 See Palmqvist 1997; Fiedel 1996 and Cattaneo et al. 1993.

187 Dealing with rhetoric, Gross states that controversy causes the search for new arguments, new evidence in the dispute: Gross 1996, xxvii. For Wagenknecht 2012, public debate had a kind of catalytic effect on academia.

dissenting voice in the media during the conference. Gibert seemed to have found what he had (apparently) been looking for: an end to the controversy through a public presentation of the arguments and the interaction between supporters and sceptics.[188] However, some unexpected critics had other ideas on how to express their disagreement.

3.4 Scientific Conferences: Much More than Debates among Colleagues

Like Leakey did 50 years earlier, in his conference, Gibert used big names from the scientific community, fossil displays, talks, media, and site visits in order to get public and political support and funding, and to convince sceptical scientists. Therefore, in these conferences, the target 'publics' multiply; they are composed of more than just scientists. The general public and politicians, as well as journalists, are also actively involved. Moreover, these new 'publics' are themselves used as 'tools' in interactions with the other sectors. For example, Leakey used the acceptance letters of great scientists to convince politicians and Gibert called on the Queen of Spain to be honorary chair and, thus, an asset to the conference. With these 'tools', conferences become a reflection of researchers' own scientific claims. Therefore, conference success means the validation of their claims, the legitimisation of their research project, and the recovery of their own scientific image and credibility.

For Gibert, this meant not only funding and permits but also a new era for his career, with recovered prestige as a scientist and with his claims validated. Somehow, the whole conference was in itself a 'tool' used by Gibert to achieve these aims within the bigger frame of the Orce controversy. Moreover, during the conference, Gibert forced opponents to go public and to give an opinion on the controversy in an environment that was very favourable to his own claims. This led Gibert to won 'three victories' (scientific, public and political).

To sum up, the 1995 Orce International Conference on Human Palaeontology appeared to amplify and concentrate several trends that defined the Orce Man controversy, scientific controversies in general, and even daily scientific practice. 'Tools' to 'convince' colleagues are always used by scientists, but the Orce conference allows a closer analysis of these 'tools' and the way they are used.

188 See more on this kind of scientific controversy closure in Machammer/Pera/Baltas 2000, 32.

In addition, this 'concentration' also revealed Gibert's and his opponents' strategies in the scientific controversy, which are no different from those used by other well-known scientists such as Louis Leakey. The Orce conference had a crucial political dimension, mainly in the relationship between scientific practice and politics and in the uses that scientists and politicians made of one another. This relationship is often hidden from stories of the Orce Man and more generally from conventional stories of scientific practice. Finally, in this analysis the mass media appear as crucial in the way that scientific knowledge is created and disseminated. All these 'strategies' and 'tools' and their political uses are only valid when they appear in the media; when the public is involved. To conclude, what Gibert aimed to achieve with the conference was presented as scientific but was only accomplished through close contact with the public, and would be of little use without it.

CHAPTER 4

End (1996–2007)

13 October 2007, Venta Micena, Orce. The site where the famous, yet almost forgotten, Orce Man was found had been closed for many years and no scientific research had been carried out there. However, on that particular day, it was full of people. Blanca Gibert, Josep's daughter, played a beautiful song on the viola. Lluís, Josep's son, spread his father's ashes on the site. Josep Gibert i Clols had just died of lymphatic cancer.

Along with Josep, the main defender of the Orce Man, the controversy also faded away. Before that, after the 1996 Conference, some of Gibert's collaborators changed side and criticized him in the public sphere. The main aim of this attack was to dismantling the conference success and Gibert's 'local power' in Orce, in order to control the remains and the site. To control both the Orce bone physically and the discourses around it was crucial for both sides of the controversy, as they allowed control and access to excavation permits and research funding.

After this attack, at the end of his life, Gibert was marginalized and ignored by his peers during a long process, in which their used different strategies to side-line him, and he also developed new ones to defend himself.

4.1 An Unexpected Attack

These three 'victories' (scientific, public, and political) did not last long. Just a month after the conference, on 6 October 1995, *El País* published an article entitled 'Palaeontologists and geologists question the existence of the Orce Man' (figure 14).[1] The piece reviewed the presentation of the Orce evidence at a scientific meeting in Madrid. As mentioned above, the Atapuerca researchers questioned the human nature of the fragments, but here also the dating of the site was also heavily criticised by participants. Alain Turq (a French scientist who collaborated with Gibert) attended the meeting, and apparently, Bienvenido Martínez-Navarro did too.[2]

1 'Paleontólogos y geólogos cuestionan la existencia del hombre de Orce', Rivera, *El País* 1995.

2 AJG-ICPMC: Letter from Alain Turq to Gibert, 11 November 1995 and Interview with Lluís Gibert 2012.

 DOI:10.1163/9789004431508_006

Paleontólogos y geólogos cuestionan la existencia del 'hombre de Orce'

Los expertos dicen que la datación no se ha hecho directamente en los yacimientos

ALICIA RIVERA, **Madrid**

En la IX Reunión Nacional sobre Cuaternario, celebrada en el Centro de Ciencias Medioambientales (del CSIC), en Madrid, los expertos de la sesión dedicada al pleistoceno han negado que se hayan realizado en el yacimiento de Orce (Granada) las debidas dataciones, han cuestionado que los fósiles hallados sean de homínido y han puesto en duda la base científica de afirmaciones realizadas por los responsables de la excavación. Los especialistas han destacado, sin embargo, la importancia del yacimiento, sobre todo por los instrumentos de piedra hallados.

El responsable de la excavación de Orce, José Gibert, afirma que para las dataciones se ha utilizado, principalmente, estudios del entorno.

"Orce es una cuenca con un gran potencial pero falta una buena documentación cronológica, sobre todo magnetoestratigrafía, que es lo más apropiado allí", afirma Alfredo Pérez, director del Instituto de Ciencias Medioambientales que ha organizado la reunión del Cuaternario. En la sesión *Hombre y medio en el Pleistoceno de la Península Ibérica*, siete especialistas abordaron el tema de Orce. En un congreso celebrado allí mismo el mes pasado, Gibert enseñó los yacimientos, fósiles e instrumentos de piedra que, afirma, tendrían en torno a 1,8 millones de años.

La datación científica de toda pieza paleontológica es imprescindible para conocer su valor y significado. "No se ha hecho el estudio geológico de los dos yacimientos Barranco León y Fuenteneuva II, donde está la industria lítica y los fósiles", continúa Pérez.

CHARO VALENZUELA

José Gibert, junto a los restos de un hipopótamo en el yacimiento Barranco León (Orce).

Algunas muestras

"No hemos estudiado estos yacimientos, como se ha dicho; lo único que hemos hecho es la magnetocronología en dos columnas estratigráficas de la cuenca de Guadix-Baza, pero están lejos de los yacimientos. Y no se pueden extrapolar los datos porque es una cuenca en que los estratos cambian abruptamente en poca distancia y es muy difícil establecer líneas de tiempo", dice Josep María Parés, jefe del laboratorio de paleomagnetismo del Instituto Jaume Almera, del CSIC, en Barcelona. "Tenemos algunas muestras de estos yacimientos pero todavía no se han acabado de analizar".

Pérez está de acuerdo: "No se puede hacer la correlación. Gibert afirma hechos que no tiene demostrados. Son hipótesis de trabajo no contrastadas".

Por su parte, Gibert, que no estaba presente en la reunión del Cuaternario, dice que sus estimaciones de edad se basan en varios estudios del entorno: "hemos utilizado el paleomagnetismo que hemos podido, la datación sigue siendo de la zona". Enumera una tesina, dos artículos y comunicaciones al congreso, y añade: "Tienen el material del yacimiento hace un año y no lo han elaborado".

En cuanto a los fósiles, José María Bermúdez de Castro, paleontólogo del Museo Nacional de Ciencias Naturales y experto en dentición, afirma que el pequeño fragmento de diente que le fue mostrado durante el congreso de Orce "no es humano". "Creo que tampoco es de homínido el humeral infantil y que no se puede reconocer el húmero", continúa. Bermúdez de Castro, uno de los co-directores de las excavaciones de Atapuerca (Burgos), afirma que sigue en duda el antiguo fragmento craneal (hallado en 1982) que ha sido considerado por muchos investigadores como de équido y defendido como humano por otros. "A partir de los fósiles presentados, el hombre de Orce no existe y las piezas tendrían en torno al millón de años", dice.

Diente

Respecto al fragmento de diente, Gibert afirma: "Dudar que es humano es muy duro. Fue un avance que hice en el congreso; ahora se me ha de exigir que haga la publicación correspondiente y estoy en ello". En cuanto a los fósiles de Orce, unas siete piezas aisladas, Gibert dice: "Para mí, y para otros, son humanos y lo hemos publicado en tres revistas".

Lo que no se ha puesto en duda es la importancia de los instrumentos de piedra hallados en Orce, unos 90. "Puede tener en torno a un millón de años, no 1,8 millones, como se ha afirmado", dice Eudald Carbonell, especialista en industria lítica. "Lo más interesante es que estas piezas indican que pudría existir presencia de homínidos allí", dice este arqueólogo de la Universidad Rovira i Virgili, de Tarragona. Pero él considera que no hay pruebas concluyentes de que los fósiles sean de homínidos. "Las afirmaciones que se han hecho no están documentadas y no se ajustan a los convenios de rigor científico internacional".

FIGURE 14 The *El País* article that quoted Spanish scientists' criticism of the Orce Man
FROM RIVERA, *EL PAÍS*, 1995

Some days later, Paul Palmqvist, another of Gibert's collaborators, sent a letter to José María Martín Delgado who was head of the *Junta*'s Department of Culture and who had attended the conference.[3] Palmqvist first stated that the findings and the conference had convinced everybody of the significance of the Orce research and the validity of the Orce Man. It seemed that the resources needed for a good interdisciplinary research team had finally arrived. Yet, according to Palmqvist, in the previous few days, "Gibert took off the mask he was wearing and showed his real aims": to monopolise all the resources for himself and his family.[4] According to Palmqvist, Gibert wanted to remove Bienvenido Martínez-Navarro, his former doctoral student, as director of the *Orce Museum*. In his place, he wanted to appoint his son, Lluís Gibert. Besides this first accusation, Palmqvist stocked the letter with some harsh descriptions of his research colleague: "Gibert monopolises and manipulates the media"; "Gibert does not excavate appropriately" and only wants to find a hominid; Gibert has "no scientific ability", Gibert 'lives off' others' intellectual production in a kind of scientific '*droit de seigneur*'; Gibert uses political parties and disputes for his own benefit; and he prefers a "front page in a newspaper to a scientific article in a journal".[5] Together with all of this, Palmqvist also requested some things from the politician. He asked Martín Delgado not to give complete control of the resources to Gibert. He suggested the creation of a commission to evaluate the Orce research project's funding and suggested Pascual Rivas, from the *Universidad de Granada*, to be part of it. Finally, he also suggested the return of all the palaeontological collections from Sabadell, where Gibert worked, to Orce. According to Palmqvist, who mainly carried out fauna research with Bienvenido Martínez, "whoever controls the fossil collections, controls, to a greater or lesser degree, the palaeontological research."[6] With regard to this, Palmqvist stated "in conversations I have had recently with [...] Dr. Jordi Agustí" (director of the *Institut de Paleonologia de Sabadell*) the latter claimed that the collections were taking up a lot of space in the *Institut*.[7]

---

3 Personal Archive Jordi Agustí: Letter from Palmqvist to Martín Delgado, 9 October 1995.

4 "despojarse Gibert de la máscara que hasta entonces tenía puesta y revelar sus verdaderos objetivos", *Ibid.*

5 "monopoliza y manipula los medios de comunicación", "no excava apropiadamente", "capacidad científica nula", "parasita", "derecho de pernada", "le gusta más una portada de periódico que un artículo científico en una revista", *Ibid.*

6 "quien controla las colecciones fósiles también controla, en mayor o menor medida, la investigación paleontológica", *Ibid.*

7 "a tenor de las conversaciones que he mantenido durante los últimos días con [...] Dr. Jordi Agustí", *Ibid.*

On 11 November 1995, the French archaeologist Alain Turq sent a long fax to Gibert in French with several complaints. Turq started by saying that he was "heartbroken" by sending this letter but "enough is enough".[8] Firstly, he complained about the excavation conditions that summer, which apparently were very poor. Then stated that he came to Orce as a director or co-director of the excavation at Fuente Nueva 3, which, in practice, he was, but then on paper he appeared merely as a collaborator. Secondly, Turq complained that he went to the Madrid meeting reported in *El País* and presented Fuente Nueva 3 dating from Gibert's data, which was strongly criticised. According to Turq, the attendees claimed that the dating had no scientific basis. Turq then complained to Gibert about these dating results and asked him "where is your scientific integrity?"[9] Finally, he also complained about the project that Gibert was about to present to the *Junta*. He said it had several problems, including being too focused on palaeoanthropology (considering the few and fragmented remains found) and also too focused on Gibert's figure. Both Palmqvist and Turq's letters, despite being addressed to different people and having different tones, reflected similar private complaints regarding Gibert's management of the Orce research.

Some months later, in February 1996, *La Vanguardia* published a double-paged article, which again presented the supposed confrontation between Orce and Atapuerca. According to this article, Gibert did not receive funding and his claims were not accepted because he "pays more attention to his intuition than to scientific rigour". Some days later, *La Vanguardia* published two brief notes entitled 'Disgusted palaeontologist' and 'Grateful palaeontologists'.[10] In the former, Gibert complained that the aforementioned article slandered him and his research. In the latter, Moyà-Solà and Meike Köhler stated that this article has done a great favour to Spanish palaeontology "putting each of us in our exact place".[11] Next month, March 1996, the Andalusian president called for early elections. IU voters heavily punished the party for its collaboration with the right-wing party (PP) and it lost seven members of parliament, which allowed the PSOE to govern without the 'clothes peg' problem.[12]

8 "narvé', 'trop c'est trop", AJG-ICPMC: Letter from Alain Turq to Gibert, 11 November 1995.

9 "Quand était-il de votre intégrite scientifique?" *Ibid.*

10 "presta más atención a su intuición que al rigor científico" Sella, *La Vanguardia* 1996 and 'Un paleontólogo indignado' Gibert, *La Vanguardia* 1996; 'Paleontólogos agradecidos' Moyà-Solà/Köhler, *La Vanguardia* 1996.

11 "situar a cada uno en su exacto lugar" Moyà-Solà/Köhler, *La Vanguardia* 1996. Köhler was also researcher working at the *Institut*. She later published an article with Moyà-Solà providing anatomical evidence against VM-0 hominid classification.

12 L.L.C. ABC 2015.

That same year, a short documentary entitled *Pedres que parlen* (Stones that talk) was broadcast on *30 minuts*, a show on the Catalan television channel TV3. This report presented Gibert's research as precarious and somehow neglected by institutions and colleagues, a kind of "scientific underground" caused by the "stigma" of the 1984 controversy. Gibert stated that he was "eager to find" the complete hominid and despite all the difficulties still had a "passion for knowledge". Eudald Carbonell, one of the co-directors of Atapuerca, also appeared on the documentary stating that within the international scientific community "nobody serious thinks that there are human remains in Orce".[13]

That summer, Paul Palmqvist and Alain Turq joined forces with Bienvenido Martínez-Navarro (figure 12). He was Gibert's former doctoral student and he considered Gibert his "scientific father". Gibert's three collaborators publicly announced their split from Gibert's team and the formation of a new independent interdisciplinary research team. The announcement was first issued at the local Orce magazine, *La Alcazaba*, then at Andalusian and national newspapers, and finally, on 16 August (during the local Orce festivals), at the Andalusian television.[14] Palmqvist, Martínez-Navarro, and Turq claimed that they were tired of Gibert's lack of funding and of his 'obsession' with the Orce Man's cranial fragment and with hominids more generally. Martínez-Navarro described the Orce hominid remains as "four miserable bones", while Palmqvist went so far as to say that Gibert was a fraudster since he had given him a simplified sketch of the Orce bone on purpose. This simplification had altered the results of the fractal analysis that he had conducted.[15]

Therefore, just a year after the publication of his article, Palmqvist changed his mind regarding the Orce fragment and even accused Gibert of fraud for having provided a false drawing to perform the fractal analysis. In 1997, the same *Journal of Human Evolution* published Palmqvist's re-evaluation of the fractal analysis.[16] This time, he used a new "accurate" and "correct" drawing of the Orce cranium's sutures provided by Salvador Moyà-Solà and "obtained from several detailed photographs", since Gibert had the original fossil. With this new and much more complex drawing, the fractal dimensions of the Orce fragment now fell within the range of the equine family (figure 15).[17]

---

13 'Pedres que parlen', "Clandestinitat científica", "ànsia trobar", "passió pel coneixement" and "Ningú seriós es creu que hi hagi restes humanes a Orce", Guàrdia/Pou 1996.

14 "padre científico", Ortuño, *Diari de Terrassa* 1987 and De León-Sotelo, *ABC* 1996b.

15 "cuatro huesos miserables", Gibert 2004, 408. For Palmqvist acusations see: De León-Sotelo, *ABC* 1996c And for this dispute in the press see: Redacción, *ABC* 1996; EFE, *La Vanguardia* 1996; De León-Sotelo, *ABC* 1996b.

16 Palmqvist 1997.

17 *Ibid.*, 85.

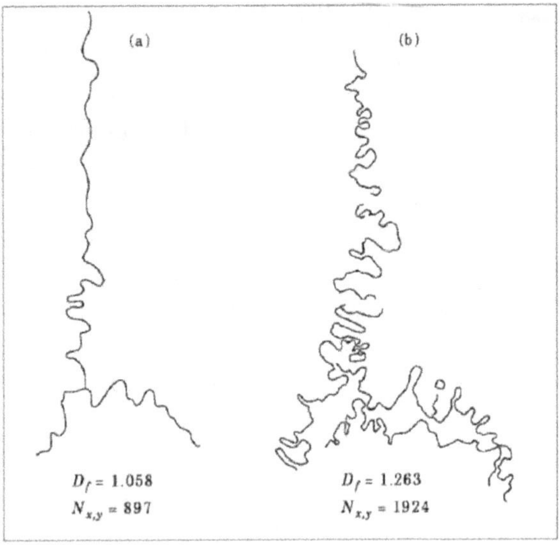

FIGURE 15 Gibert's (left), and Moyà-Solà's (right) drawings of VM-O's sutures
PALMQVIST, 1997, 85

Finally, in their announcement, Palmqvist, Martínez-Navarro and Turq concluded that the Conference attendants were just polite by not criticising Gibert during the Conference but most of them did not believe Gibert's claims.[18] In his 2005 popularization book, Agustí stated that during the Conference, to which he did not assisted, stone tools and specially its dating were received with scepticism by the assistants.[19]

In his response to the split, Gibert considered these statements far worse than all the criticism he had received so far during the controversy.[20] For him, it was one thing to change opinions in science and quite another and much more serious thing to accuse somebody of fraud. The public presence of the controversy had immediate consequences when the Orce Town Council dismissed Martínez-Navarro as director of the *Orce Museum*. According to the newspaper *ABC*, the reason from the dismissal was that, "these kinds of statements [referring to Martínez-Navarro's appearance on television] seriously damage the image of this town."[21] The mayor of Orce, Leandro Torres, first stated that there

18 De León-Sotelo, *ABC* 1996c.

19 Agustí/Lordkipanidze 2005.

20 Gibert/Gibert 2000, 230–231 and Gibert 2004, 407–409.

21 "este tipo de declaraciones perjudican enormemente la imagen de este pueblo", De León-Sotelo, *ABC* 1996a.

was no problem with Martínez-Navarro thinking that the bone was not a hominid, but that television was not the correct medium to express this opinion. However, later, the mayor added, "here we have a museum, and the cranium is the star".[22]

Martínez-Navarro's dismissal reveals how the town needed the Orce Man to be accepted among the scientific community in order to use it as a tourist attraction. The Orce Man remain was always present in Orce's town and museum advertising.[23] At the same time, Palmqvist and Martínez-Navarro created a new research team that presented its candidacy for conducting research in Orce that summer of 1996. This new team, according to the newspapers, wanted to forget about the eternal discussion on the hominid presence in Orce and focus their efforts on the palaeontological study of the fauna and the ecological environment of the Orce Basin over one million years ago.[24] They wanted to forget about the Orce bone fragment, but first they had to publicly destroy it. According to *ABC*, in the summer of 1996, neither Gibert's team nor the new research team was able to conduct research in Venta Micena.[25]

Sometime later, the team's split took shape with the publication of some scientific articles. In December 1996, Martínez-Navarro, Palmqvist, Turq, and Agustí published an article reviewing all the research carried out in Orce.[26] A year later, the prestigious *Journal of Human Evolution* published two articles in the same issue on the Orce Man. In the first of these articles, as already said, Paul Palmqvist revised his fractal analysis with a new VM-0 sketch: the bone was no longer hominid (figure 15). In the second paper ('The Orce Skull: Anatomy of a Mistake'), Meike Köhler and Salvador Moyà-Solà, one of the piece's discoverers, looked at the Orce bone's anatomical features and concluded that it was "just a common and ordinary horse".[27] After some years (since 1987) without publicly criticising Gibert or the Orce Man, Agustí and Moyà-Solà reappeared to join perhaps the last scientific and public attack on the Orce Man. Of course, the publication of these articles was reported in Spanish newspapers. They noted that new "crushing evidence" showed that the Orce Man was not a man anymore.[28] Journalists also highlighted that the articles had been

22 "Aquí tenemos un museo, y el cráneo es la estrella", *Ibid.*

23 AJG-ICPMC: Report of the Museo Municipal de Paleontologia, Prehistoria y Arqueologia, 'José Gibert', n.d. Flyer 'Orce, the Cradle of European Humankind', around 1992; Personal Archive Jordi Agustí: Flayer 'Orce, Cuna de la Humanidad Europea', around 1992.

24 De León-Sotelo, *ABC* 1996c.

25 De León-Sotelo, *ABC* 1996c.

26 Turq et al. 1996.

27 Palmqvist 1997 and Moyà-Solà/Köhler 1997.

28 "pruebas demoledoras", Redacción, *La Vanguardia* 1997.

published in a prestigious international journal. Both *El País* and *La Vanguardia* reproduced exactly the same sentence: the Orce Man was "a common and ordinary horse".[29]

The way that the three 'dissenters' (Palmqvist, Martínez-Navarro and Turq) presented their new opinions, first in private letters to politicians and Gibert and then in public (from the Orce magazine to the international research journals, from local to global, from general to specialised) reveals their aim to get rid of the two elements analysed: Gibert's 'power' in the local context of Orce and the three 'victories' (scientific, public, and political) won during the conference. The attack was not only directed at Gibert's own persona, but also at the famous bone fragment known as the Orce Man. This way, the three 'dissenters' wanted to close down the controversy and move forward by destroying Gibert's credibility and, at the same time, destroying the validity of the Orce Man or the other way round. This also gave them free rein to assemble a new research team to perform research in Orce and to apply for permits and funding.

While all of this was happening, Gibert was taking advantage of the benefits of the conference. He sent letters to several researchers in order to get further support for his research. By communicating with scientists, he was aiming to build an international team to study the Orce area. He also tried to get some of the Orce research published in top-ranking journals such as *Science* and *Nature*.[30] Yet in *Science*, Gibert's paper was first given "a lower priority rating" and later, due to several issues mainly about the originality of the material, the *Science* editor stated that "this is a potentially very exciting paper, but we feel that much additional clarification is needed before we can consider sending it to review."[31] Sometime later, when the paper was finally sent to review, the reviewers considered that the material presented was not enough "to justify the conclusions", nor was the uncertain dating, and was finally rejected.[32] In Gibert's archive, he kept several draft copies of this paper. In the first versions of it, Palmqvist, Martínez-Navarro, and Turq were among the co-authors. Then there are versions with a line through the three names, and finally, versions in

29 "Un caballo común y ordinario", Ruiz de Elvira, *El País* 1997.

30 For instance: AJG-ICPMC: Letter from Derek Roe to Josep Gibert, 16 October 1995; Letter from Gibert to Derek Roe, 30 October 1995; Letter from Gibert to Jaques Tixier, 24 November 1995; Letter from Gibert to Vera Eisenmann, 15 October 1996; Letter from Gibert to Ekaterina Bouliguina, 2 January 1997; Letter from Gibert to Derek Roe, 3 March 1997.

31 AJG-ICPMC: Letter from R. Brooks Hanson, *Science* Editor, to Gibert, 18 June 1996.

32 AJG-ICPMC: Letter from R. Brooks Hanson to Gibert, 25 April 1997.

which their names no longer appear.[33] Also during this period, when Gibert and Orce were again at the height of their popularity after the conference, he signed another contract with the publishing house Ariel for a popular science book, which was provisionally titled *The Dawn of Men*.[34] This book was never written.

During the new period of controversy, Gibert also had international support from two researchers that attended the conference. Phillip Tobias and Jerold Lowenstein published two articles in scientific journals praising Gibert and above all the research carried out in Orce. In *Human Evolution*, Tobias stated that the hominid bones found in Orce seemed human and together with the stone tools "provide evidence for the probable presence of hominids in Venta Micena".[35] In *Current Anthropology*, Lowenstein wrote a brief piece in which he compared Orce with Olduvai and stated that the conference "revealed hominid presence" in Orce. He also reviewed the sites and findings, his immunological research, and the evidence that pointed to the human nature of VM-0. This piece also commented on some of the conference presentations related to the Orce research and finished by stating that "Orce is now of importance not only to Spain or to Europe but to the whole world."[36]

At the same time, Gibert was also sending letters to politicians, specially to Catalan politicians that were not as much involved in the Conference. Using the success of the meeting, Gibert asked for support to the head of culture of the *Diputació de Barcelona*, Joan Francesc Marco (PSC-PSOE) and in a couple of letters to Narcís Serra (PSC-PSOE), former mayor of Barcelona, former minister of defence and former vice-president of the Spanish Government, and at that time, head of the PSC-PSOE.[37] These letters show once again the close relation between the Orce researchers and politics and the use of the Conference in Gibert's quest for funding. In just a year, everything had changed many times. After the Conference, everything seemed positive for Gibert; the new attack put him again in the bad position of having to fight for his scientific reliability, especially in front of politicians, who were those who, at the end, had the last word in the funding giving processes.

33 AJG-ICPMC: Draft copies box.

34 AJG-ICPMC: Letter from Asunción Hernández to Josep Gibert with the publishing contract with Ariel attached, 5 June 1996.

35 Tobias 1998, 93.

36 Zihlman/Lowenstein 1996, 697.

37 AJG-ICPMC: Letter from Joan Francesc Marco to Gibert, 7 March 1996; Letter from Gibert to Narcís Serra, 15 April 1996; Letter from Gibert to Narcís Serra, 19 December 1996.

4.2 Control of the Remains Means Control of the Research

After the 1995 conference and the controversial summer that followed it, in 1997 and 1998 permits were denied to both Gibert's team and the new research team led by Palmqvist and Martínez-Navarro.[38] As already said, a month after the conference, Palmqvist wrote a letter to José María Martín Delgado, director of the *Junta de Andalucía*'s Department of Culture.[39] In it, among very harsh criticisms of Gibert and his research, he suggested the return of all of the palaeontological collections from Sabadell to Orce. In the summer of 1996, the idea to return the bones was taken up by the mayor of Orce, Leandro Torres, who claimed that more than 4,000 fossils from the early excavations in Venta Micena that were stored in Sabadell (including the famous VM-0 cranial fragment) should be transferred back to Orce.[40] For the mayor, with this return they "hoped to put an end to the problem of disputes over the human classification of the Orce Man and to try to ensure that the controversy does not affect the rest of the town's archaeological wealth."[41] For Torres, it seemed that there was some "social alarm" in Orce after Palmqvist, Martínez-Navarro, and Turq's announcement and he felt that this "has to end," and therefore, "the cranial fragment [...] must be made available to the scientific community."[42] Torres even threatened to go to court as "ownership of a discovery is not everlasting".[43] For the Orce mayor, to have the bone in Orce meant also to control the discourses, at least the local discourses around it. When it arrived to the town, the Orce Man was exhibited as a very important hominid remain in Orce's local museum. This was a local way to end the problems, from then on, at least in Orce; the bone was not controversial anymore. These claims became effective and the *Junta de Andalucía* requested the return of the fossils from the *Institut de Paleontologia de Sabadell*.[44] Isidro Toro, then head of the *Junta*'s Department of Archaeology, travelled to Sabadell to organise the transfer of the bones.[45]

---

38 Troyano, *ABC* 1997 and F. R. A., *ABC* 1998.

39 Personal Archive Jordi Agustí: Letter from Palmqvist to Martín Delgado, 9 October 1995.

40 De León-Sotelo, *ABC* 1996d.

41 "Con esta medida pretendemos acabar con el problema de las disputas sobre la humanidad del hombre de Orce e intentar que esa polémica no afecte al resto de la riqueza arqueológica del municipio", EFE, *La Vanguardia* 1996.

42 "alarma social", "con eso hay que terminar", "el fragmento de craneal debe ser puesto a disposición de la comunidad científica", De León-Sotelo, *ABC* 1996d.

43 "la propiedad sobre un hallazgo no es eterna", De León-Sotelo, *ABC* 1996e.

44 De León-Sotelo, *ABC* 1996e and Gómez, *El Ideal de Granada* 1997.

45 Gibert 2004, 99.

Gibert's response to the return of the palaeontological remains to Orce is very interesting. In retrospect, in his 2004 popular science book, Gibert stated that the return of the bones to Orce was an honour, a source of "great satisfaction" for him.[46] Yet, according to newspapers at that time, Gibert had a quite different response. First in 1996, *ABC* reported that Gibert had said that VM-0 "is mine and will go with me wherever I go".[47] Gibert also asserted that all the remains held in Sabadell, including the famous cranial fragment, were found before 1985 (when the *Junta de Andalucía* took all legal control over archaeological remains), so they were being held legally in Sabadell.[48] By the beginning of 1997, for Gibert, the transfer of the bones could only be allowed if it was guaranteed that they would be properly stored and classified.[49] Gibert also stated that what he was about to hand over were not only fossils but also "21 years of my work", often carried out in very poor economic circumstances.[50]

Finally, the remains were returned to Orce after two ceremonies, one in Granada and another in Orce, with the presence of the *Junta*'s head of culture, Carmen Calvo. Gibert brought the most valuable remains in a metal suitcase, mainly the bones claimed to be human, including, of course, the so-called Orce Man, and some of the stone tools. According to Gibert, these remains were wrapped in "cloths that highlighted their importance" and were shown to the political authorities and the media.[51] Again, Gibert seized the opportunity to request new excavation permits and funding. All of this was reported on in the press, and even made the front page of the Andalusian newspaper *El Ideal de Granada,* which showed Gibert together with Carmen Calvo, the *Junta* head of culture, that years later also became Spain's government vice-president.[52] The Orce Town Council announced the improvement of the local museum (the *Museo de Paleontologia y Prehistoria de Orce*) to hold the original cranial fragment of the Orce Man and the other remains.[53] Pascual Rivas, head of palaeontology in the *Universidad de Granada*, and a member of the *Junta*'s Department of Archaeology, was one of the only dissenting voices, stating in the papers that a museum cannot be built around a "questionable fossil".[54]

---

46 "Gran satisfacción", *Ibid.*

47 "El cráneo es mío e irá conmigo donde yo vaya", De León-Sotelo, *ABC* 1996e.

48 *Ibid.*

49 EFE, *ABC* 1997.

50 "21 años de mi trabajo", Canal Sur 1997.

51 "paños que resaltaban su importancia", Gibert 2004, 101–102; F. R. A., *ABC* 1998a; F. R. A., *ABC* 1998b; Gómez, *El Ideal de Granada* 1998.

52 Gibert 2004, 102–103 and Gómez, *El Ideal de Granada* 1998.

53 Troyano, *ABC* 1998b.

54 "fóssil dudoso", Titos, *El País* 1999.

This contradiction between a sense of owning the remains while simultaneously being proud of their return to the place where they were found is quite common in palaeoanthropology. Hominid remains are very scarce and hard to find. Palaeoanthropologists often limit access to remains, which creates confrontations between researchers over priority and publication.[55] As said, hominid findings offer great professional opportunities. Due to these circumstances, researchers often have a sense of 'possession' of the bones and even see themselves as 'fathers' of the fossils they find.[56] Hominid fossils thus become world heritage treasures and, at the same time, objects of science 'owned' by particular scientists. Like Gibert, Donald Johanson described how the return of Lucy to Ethiopia was very important and had great symbolic significance. Yet, at the same time, the American researcher described his negative feelings about becoming detached from the remains and acknowledged that he was "the owner of an unparalleled collection of bones".[57] To sum up, both Orce's desire to have the important remains again and Gibert's 'ownership' of the Orce Man combined with his mixed feelings about its return to Orce are no different from other cases in international palaeoanthropology.

Still, the desire to control Orce's remains was not exclusive to Gibert. As Palmqvist's letter, Rivas critiques to the *Orce Museum* or Toro's trip to Sabadell show, the scientists from the 'other side' also wanted to get their hands on the Venta Micena collection. Yet, while Gibert's position is typical in palaeoanthropology: Toro, Palmqvist, Martínez-Navarro, and even Rivas' is not. They wanted to have control over the Orce Man even when they denied the hominid origin of the cranial fragment as well as the hominid presence in Venta Micena. So, why did they want to control a bone fragment recognised as an ancient donkey remain? Why did they want to have 4,000 palaeontological specimens from Venta Micena, a site in which further research was being blocked by Rivas and Toro? One first obvious answer is that Palmqvist and Martínez-Navarro were studying Orce's fauna and, for sure, the Venta Micena remains would be very useful to them. Yet, behind this more scientific aim, there was also the aim to take power away from Gibert by taking the disputed hominid remains away from him, most importantly the well-known Orce Man. By the end of the 1990s, this bone was still the flagship of Gibert's research.

55 Gibbons 2002.

56 In his memoir, Phillip Tobias devoted a whole section to scientists' inclination towards feeling ownership over bones, mostly using Dart as an example, Tobias 2005, 228–231 and 237.

57 Johanson/Maitland 1982, 206 and 332.

To sum up, the Andalusian researchers not only wanted to control precious scientific objects that could lead to important publications, but also 'famous' scientific objects that could bring Gibert again and again into the media. Therefore, in this case, the purely scientific relevance of the remains was not the main motivation for scientists to want to claim and control them; rather, it was their public relevance, especially that of VM-0, the so-called Orce Man. Thus, despite Martínez-Navarro, Palmqvist, and Turq's discourse of forgetting about the hominid to perform much more interdisciplinary research in Orce (with studies in fauna and ecology); in the end, control over hominid remains was a very relevant issue for their team. During the following years, this team used fauna discoveries to reassert the claim that the Orce area held the remains of the 'First European', a claim that essentially was no different from Gibert's. As the newspaper *ABC* announced, 'The remains of an elephant reinforce the hypothesis that Orce did have human presence'.[58]

In March 1997, *El Ideal* announced the further publication of a scientific article also in *Journal of Human Evolution* by Paul Palmqvist in which he re-evaluated the fractal analysis of the bone and arrived to the conclusion that the bone was not human, but most probably of an animal of the horse family.[59] While Palmqvist paper was through the peer-review process, the JHE editor, Lesille Aiello, who was invited though did not attended to the Conference, gave the opportunity to Gibert and his team to answer Palmqvist' conclusions with another article. The original idea was to publish both articles together, yet Gibert's delay in the publication caused the Palmqvist paper to be published first.[60]

The *El Ideal* article mentioned Aiello's and the *Journal of Human Evolution* acceptance of Palmqvist article to be published, and therefore somehow supporting his claims. According to the text, the journalist, J. E. Gómez, had read Aiello's letter of acceptance to Palmqvist. After this apparent support of the international journal to the rivals of the Orce Man, the Orce mayor, Leandro Torres, sent a letter to Aiello showing his concern about the claims made by Palmqvist in *El Ideal* and asking Aiello to confirm that Palmqvist claims were true and that the journal supported them, as the newspaper had announced.[61] Next day, Aiello answered the Orce mayor that the paper was not yet published

58 'Los restos de un elefante refuerzan la hipótesis de que Orce tuvo presencia humana', Redacción, *ABC* 2001.

59 Gómez, J.E. *El Ideal de Granada* 1997.

60 Letter from Gibert to Aiello, 22 March 1997; Letter from Aiello to Josep Gibert, 25 March 1997; Letter from Gibert to Aiello, 25 March 1997; Letter from Gibert to Aiello, 1 April 1997. Gibert's article was finally published in 1998: Gibert et al. 1998 and S. C., *ABC* 1998.

61 Letter from Leandro Torres to Aiello, 26 March 1997.

and that it was at the moment through the process of peer-review and also explained the opportunity that the journal gave to Gibert to elaborate a response.[62] This is yet another example to how close was, during the Orce man long controversy, the relations between politicians, journalists and scientists.

These close relations are, on the one hand, crucial for understanding the key issues raised. They also highlight the importance of the geographies of science. Aiello from London, Gibert from Sabadell, Palmqvist from Málaga (where he was professor at the University), the journalist from Granada, and Torres from the small town of Orce defended their own aims and interests but also represented certain geographical stances that have already been discussed: mainly, the fluid and to a large extent inadequate notions of 'centre' and 'periphery'. The 'centre' position, be it Aiello, de Lumley, or even the Atapuerca researchers, was used in this story by the allegedly 'peripheral' actors in a clearly double-sided discourse. On one hand, as bearers of prestige, impressing the Spanish public and politicians, and, on the other hand, as 'threats' from abroad that came to Spain, or to Orce, to spoil 'our' archaeological treasures. These discourses conceal a strategy of portraying oneself in the periphery for the purposes of legitimisation and funding. In addition, these same discourses contribute to reaffirming and perpetuating these questionable notions of 'centre' and 'periphery'.

Likewise, the centre also pursued its own interests, mainly to 'control' the research carried out in Orce or to maintain prestigious positions and reputations in the scientific sphere but also, and very importantly, in the public sphere too. This becomes very clear by analysing the Atapuerca co-directors' strategic moves each time that the Orce Man stood out in the media. The way that the three co-directors appeared and the timing of their appearances (during and after the 1995 Orce conference, when the Orce Man appeared in all the media) clearly reveal their aim to refute publicly Gibert's claims and even destroy his credibility in order to control palaeoanthropological research in Spain.

From the double 'peripheral' situation of the small town of Orce, local politicians said that they wanted to leave research up to the most qualified scientists, but when the Orce Man was called into question, the Town Council made the moves it could make (whether dismissing Martínez-Navarro, claiming the remains or sending a letter to Aiello) to try to keep the Orce Man alive, at least at the local level. The Council also had the aim of transforming the town into a prehistoric tourist attraction and therefore providing economic opportunities for the villagers. In Orce archaeological research, and more specifically

62 Letter from Leslie Aiello to Leandro Torres, 27 March 1997.

the Orce Man, represented the improvement of their living standards. In this situation, Gibert, as the main advocate of the Orce hominid, somehow became Orce's local representative of scientific research. Orce became his refuge, where the Orce Man was presented without doubt not only as a hominid, but also as the 'First European'.

All of this leads us to reconsider the traditional concepts of 'centre' and 'periphery' in scientific activity. In this book, I have tried to understand how these are not fixed concepts but can vary and are often positions created and perpetuated by historical actors themselves in their discourses in efforts to fulfil their own agendas. The analysis of the geographies of science should then not consider these prior concepts of 'centre' and 'periphery' of knowledge but take into account the historical actors' relations with other actors, the use of these relations in their discourses, and the significance of these relations in the development of scientific practice.

∴

As mentioned earlier, during this period the *Junta* created an *International Commission* to control research in Orce, again following one of Palmqvist's recommendations.[63] The main aim of the Commission was to review the scientific projects presented for the Orce area. It was headed by Pascual Rivas and included Claude Guérin, a French palaeontologist from the *Université Claude Bernard* (Lyon), Gerhard Bosinski, a German palaeontologist, a friend of Gibert's, and an Orce conference attendee, and some other Spanish scientists, some of them also Orce conference attendees.[64] Jordi Agustí, Isidro Toro, and Josep Gibert were in charge of the scientific project with participants from both Gibert's team and Martínez-Navarro and Palmqvist's new team.[65] In this context, Gibert again looked for support for his views. This time, Raimon Obiols, president of the Catalan PSC-PSOE, received a letter from Gibert asking him to talk with Manuel Chaves (member the Andalusian PSOE and then president of the *Junta de Andalucía*) about partiality in the creation of the Commission, which granted excavation permits in Orce. In short, Gibert asked Obiols "to contribute to avoiding serious injustices being committed".[66] Once

63 Personal Archive Jordi Agustí: Letter from Palmqvist to Martín Delgado, 9 October 1995 and Gibert 2004, 102.

64 For Guérin: Gutiérrez/Karlin/ Faure 2016. For Bosinski see: Wikipedia Contributors 'Gerhard Bosinski'.

65 Gibert 2004, 102–103.

66 "contribuir a que no se cometan graves injusticias", AJG-ICPMC: Letter from Gibert to Raimon Obiols, 28 January 1997.

again, he used direct contact with politicians in order to champion his cause regarding the Orce research.

During this period, on 24 April 1998, the Orce issue was debated in the Andalusian Parliament. This time, Javier Torres Vela (PSOE), president of the Parliament, gave the floor to Carolina González Vigo (PP), Member of Parliament from the right-wing PP, who corresponded with Gibert. González Vigo wanted to ask a question requesting an oral answer from the *Junta*'s head of culture, Carmen Calvo.[67] The PP member of parliament took the stand to ask "what criteria and what specific projects does the Department of Culture have in relation to Orce's archaeological sites and in relation to the granting of excavation permits to continue research on them?"[68] Vigo also asked the *Junta*'s head of culture about the plans for the Orce research and the granting of permits.[69] The head of culture, Carmen Calvo, responded that the recently created Commission, formed by several international experts, had recommended the denial of the permits.[70] Soon afterwards, Gibert sent a letter to González Vigo, in which he stated that not all of the experts from the Commission had recommended the denial and he attached a letter from Bosinski to prove it.[71] This is, at least, the third time that the Orce issue was bought to the Andalusian parliament. It is unknown if the Orce issue was brought once again to parliament, but this situation once more reveals the political interest in the issue.

Finally, excavation permits where granted to Gibert for the summer of 1999.[72] Yet, they were granted for Barranco Leon and Fuente Nueva 3, not Venta Micena, the site where the controversial cranium was found, and which remained closed.[73] In fact, apart from a brief permit for the conference, Venta Micena had been closed since 1990; almost 10 years.

To sum up, 'control' is the key issue in this section. With the new controversy, the different parties' desires to control the archaeological remains, the sites, and the excavation permits show up very clearly. These different parties, Gibert and Palmqvist and the new team, but also the mayor of Orce or the *Junta* politicians, made several moves and developed strategies around the Orce Man's cranial fragment in a kind of 'fight for control', although with different

67 "con ruego de respuesta oral", AJG-ICPMC: Debate minutes, Sesión Plenaria 23 April 1998.

68 "¿qué criterios mantiene y qué proyectos concretos tiene la Consejería de Cultura en cuanto a los yacimientos arqueológicos de Orce así como a la concesión de permisos para seguir investigando en ellos?", AJG-ICPMC: Debate minutes, Sesión Plenaria 23 April 1998.

69 AJG-ICPMC: Debate minutes, Sesión Plenaria 23 April 1998.

70 *Ibid.*

71 AJG-ICPMC: Letter from Gibert to González Vigo, 1 June 1998.

72 Troyano, R., *ABC* 1999.

73 Ruiz Antón, *ABC* 1999.

aims. On the one hand, the Orce Man was Gibert's main scientific achievement, representing his own credibility and standing as a member of the scientific community. On the other hand, for Palmqvist, Martínez-Navarro, and the other members or supporters of their team and their claims, the Orce Man was not in fact a man, but they needed to control it in order to control the research and the claims made in the area.

This team's strategy was to erase and forget the hominid claims, specifically the Orce Man, in order to have a fresh start with a new research project. To be clear that the Orce Man was not a man was as important for Palmqvist and the others as it was to 'control' what happened with it. To achieve this control and discourse over it, the Orce Man could not be in Gibert's hands and must be controlled from Andalusia. With a discourse of wanting to end scientific disagreement and provide access to the remains for the scientific community, the Orce authorities achieved the return of the remain to the town, where it was also a hominid. VM-0 has been always a prominent image in the Orce town tourist promotion. According to Lluís Gibert, around 2012, the original Orce Man fragment was brought by the Orce Town Council to a tourism trade fair in Madrid and presented as a tourist attraction.[74]

In 2015, a new museum was open in Orce. Despite a lot of research has been performed in the last years, and a new hominid remain had been found, the Orce Man original fragment still had a prominent position as a bone 'attributed' to a hominid (figure 16). The Orce mayor José Ramón Martínez Olivares discourse in the inauguration stated that palaeontological and archaeological research was a very useful way to attract tourism to the area and therefore generate richness and employment to the town inhabitants.[75] For the *Junta* politicians, mainly the head of culture, the Orce remains represented an opportunity to appear before Andalusians promoting culture and scientific research. But, and maybe more importantly, the pompous ceremony of the delivery of the bones from Gibert to the *Junta* shows how these politicians also presented themselves as having achieved the return of important 'Andalusian' heritage to Andalusian soil.

All of these different representations from different actors in the story show, on the one hand, the extent to which the Orce Man fragment had become a famous and important scientific object both in public and scientific arenas. More than fifteen years after the discovery, the fragment was still commented

74 Personal communication with Lluís Gibert. Martínez-Navarro 2012, 16 also states that the fragment still is a significant part of Orce's cultural heritage despite the years passed since its discovery and the subsequent controversy.

75 Utrera, *El Ideal de Granada* 2015.

FIGURE 16 Orce inhabitants looking at VM-0 during the 2015 inauguration
PHOTO CLARA FLORENSA

on, disputed, and desired by different parties. On the other hand, this situation shows the different faces that a simple scientific object can have. For both the institutions and the media in Andalusia, the Orce remains were much more than just a simple scientific object. The Orce Man was not anymore just a natural object, but it was, at the same time a way to attract tourists, a good topic for a news piece or an asset in the political fight; a hominid or a donkey, the Orce fragment was a public object with different interpretations and uses.

4.3 The Process of Isolation

After the creation of the new research group led by Palmqvist and Martínez-Navarro, Gibert and his team suffered a process of isolation and separation from the Orce research. For a better understanding of this process, this section will review the excavations performed in Orce from 1999 to 2005.

In the summers of 1999 and 2000 everybody worked in a shared Orce research project entitled 'Geology, Palaeontology, Palaeoecology, and Archaeology in the Guadix-Baza Basin during the Plio-Pleistocene', which was supervised by the *International Commission*.[76] In 2000, this project was directed, according to *El País*, by Gibert, Martínez-Navarro, Jordi Agustí, and Isidro Toro.[77] Yet, in November of the same year, Toro, Agustí, and Martínez-Navarro organised a meeting in Granada to present an exhibition on the 'First Europeans'. Gibert was not invited to the meeting and the disputed Orce remains were not exhibited. Newspapers highlighted Gibert's absence but mainly followed the organisers' discourse that "the debate has clouded [...] the real significance of the Orce sites, which do not have hominid remains but which do show, through the discovery of lithic industries, the presence of man's ancestors."[78] Gerhard Bosinski, member of the *International Commission* of the Orce research project; Juan Luis Arsuaga, one of the co-directors of the Atapuerca research team; Clark Howell, a prestigious American palaeoanthropologist and Orce conference attendee; David Lordkipanize, head of the research conducted at Dmanisi; and Henry de Lumley all attended the meeting in Granada.[79] Just after this scientific meeting, another meeting was held by the *International Commission*. Gibert did attend this meeting, as did most of the participants from the 'First Europeans' meeting, even if they were not officially part of the Commission.[80] Juan Luis Arsuaga, from Atapuerca's research team, was appointed as a new member of the Commission in that meeting.[81] In an interview a week after the meeting, Arsuaga stated that he "never believed it [the Orce Man] was a human cranium", and that he had no rivalry with Orce but only with Gibert who was still defending the disputed cranial fragment.[82] During the meeting, Gibert asked the Commission about the possibility of working in Venta Micena. The answer was that Venta Micena had to be "appraised" by the Commission before being excavated.[83]

76 'Geología, paleontología, paleoecología y arqueología de la cuenca Guadix-Baza durante el pliopleistoceno', Arias, *El País* 2000a.
77 Arias, *El País* 2000b.
78 "El debate ha empañado la verdadera importancia de los yacimientos de Orce, que no ofrecen restos humanos, pero si demuestran, mediante el descubrimiento de industrias líticas, la presencia de antecesores del hombre", Arias, *El País* 2000c.
79 *Ibid.*, and Gibert 2004, 104.
80 Gibert 2004, 104.
81 Arias, *El País* 2000d.
82 "jamás creí que fuera un cráneo humano", *Ibid.*
83 "puesta en valor", Gibert 2004, 104.

In 2001, the Orce research project again conducted excavations in Barranco León and Fuente Nueva 3, and also in a new site, Barranco del Paso, with Gibert in charge. According to him, this was an interesting but not very productive site.[84] The Barranco León excavations, led by Toro, Martínez-Navarro, and Turq (according to the media), were reported after the discovery of spectacular faunal findings, like the remains of a giant deer and an elephant pelvis.[85] As already indicated, these reports were accompanied by a discourse of Orce as the first place of hominid occupation in Europe, despite the authors not recognising any of the supposed hominid remains found by Gibert's team over the years. Beyond stone tools, they stated that the faunal remains also supported this claim, as they were very similar to those found in Dmanisi, Georgia, an older site that had been recognised as the first hominid site outside of Africa.[86] It seemed that the same fauna had migrated from Africa to Georgia and to Orce, including hominids. Martínez-Navarro even stated that they "are not in a rush to find hominid remains. It is certain that they are there, and that they will appear, but there is no need to find them to confirm that the oldest human presence in Western Europe is in Orce."[87] In this sense, in 2002 Martínez-Navarro published a scientific article that reclassified VM-0. For him, the famous Orce Man bone was not a hominid, but it was neither an equid. Instead, for Bienvenido Martínez-Navarro, who had collaborated first with Gibert and later left his team, VM-0 was now the frontal bone of a female of a large ruminant without antlers.[88] This new attribution shows Martínez-Navarro's will to, through a scientific publication, be detached from past controversies and to start a new research period without being always related to the Orce Man.

In 2002, excavation permits for Venta Micena were denied both to Gibert and Martínez-Navarro.[89] According to *El País*, Gibert's project was rejected due to "scientific and administrative reasons" and to a negative report from the *International Commission*.[90] The *Junta de Andalucía*'s head of culture even described Gibert's project as having "low scientific quality" and Gibert as someone "that researches with very little scientific criteria."[91] Finally, *El País* also

84 *Ibid.*, 105.

85 EFE, *El País* 2001; A. V. G., *El País* 2001; Redacción, *ABC* 2001 and Martínez, *ABC* 2001.

86 A. V. G., *El País* 2001.

87 "No hay prisa por encontrarlos. Seguro que los hay y que aparecerán, pero no es necesario encontrar amainados para confirmar que en Orce se encuentra la presencia humana más antigua de Europa Occidental", Martínez, *ABC* 2001.

88 Martínez-Navarro 2002 and Martínez-Navarro 2012.

89 Quero, *El País* 2002.

90 "Razones científicas y administrativas", *Ibid.*

91 "baja calidad científica", Quero, *El País* 2002; "que investiga con poco criterio científico", Arias, *El País* 2002a, and Arias, *El País* 2002b.

indicated that the three co-directors of the Orce excavations in Fuente Nueva 3 and Barranco León were Jordi Agustí, Bienvenido Martínez-Navarro, and Isidro Toro.[92] Gibert in the end had the opportunity to carry out excavations in Fuente Nueva 1, another secondary site near Fuente Nueva 3.[93]

The following year, Gibert's excavation request was denied again.[94] Instead, an excavation permit for Venta Micena was granted to the *Institut de Paleontologia de Sabadell*. Jordi Agustí, still director of the *Institut*, also became the director of the excavations and decided that Gibert and Martínez-Navarro would both conduct excavations in two different parts of the site.[95] According to Gibert, this new strategy was a disaster, and it did not work.[96] On 1 August 2003, the excavations officially ended. Agustí, as director, gave the so-called 'excavation book' to Isidro Toro, the archaeologist from the *Junta*. This is the legally required official book in which the work carried out and the remains removed are recorded.[97]

That same day, a scientific monograph on the recent excavations was presented in Orce with the attendance of the *Junta*'s head of culture and other politicians.[98] This work was entitled *The Lower Pleistocene in Barranco León and Fuente Nueva 3, Orce (Granada). A Scientific Memoir of the Excavations 1999–2002*, and was signed by Jordi Agustí, Isidro Toro, and Bienvenido Martínez-Navarro.[99] On 4 August, Gibert apparently received a new official excavation book from the *Junta*, which allowed him to continue digging in Venta Micena.[100] Some days later, the Andalusian police appeared in Venta Micena where Gibert's team were still excavating. Apparently, the excavation was illegal and therefore the police filed an official report. The next day, Gibert handed over the second official book signalling the end of the excavation, but, according to Agustí, it seems he did not have an official permit to continue excavations.[101]

92 Quero, *El País* 2002.

93 Redacción, *ABC* 2002, and Gibert 2004, 105.

94 Quero, *El País* 2003.

95 Quero, *El País* 2003 and Gibert 2004, 105.

96 Gibert 2004, 105.

97 Trillo/Muñoz, *Granada Hoy* 2007; Interview with Jordi Agustí 2012 and Interview with Lluís Gibert 2012.

98 Redacción, *ABC* 2003 and Trillo/Muñoz, *Granada Hoy* 2007.

99 *El Pleistoceno Inferior de Barranco León y Fuente Nueva 3, Orce (Granada). Memoria Científica Campañas 1999–2002*, Redacción, *ABC* 2003.

100 Trillo/Muñoz, *Granada Hoy* 2007.

101 Interview with Agustí 2012; Interview with Lluís Gibert 2012; Trillo/Muñoz, *Granada Hoy* 2007.

In December 2004, over a year later, the *Junta de Andalucía* imposed a fine on Gibert of 60,000 euros and prohibited him from being head of any archaeological project as a result of the police report.[102] According to Gibert, this was the biggest penalty ever imposed in Spain on a palaeontologist or archaeologist, even bigger than those imposed on people that damage or steal from archaeological sites.[103] In January 2005, Gibert presented an appeal to the *Junta*. He said that since the authorities had given him a new excavation book he thought everything was in order, despite Agustí, head of the excavations, not being there. He also added that due to Agustí's delay, the excavation began late so they did not have enough days to carry out all the work. Gibert also claimed that on 1 August quite a few bones, more than 700, were exposed and needed to be removed before the end of the excavation.[104] In March, the local newspaper *Granada Hoy* presented the case on the cover and devoted two full pages to the issue.[105] In July 2005, the *Junta* confirmed Gibert's penalty, and he appealed against it again. In the end, Gibert did not assumed the whole fine, but, according to Lluís Gibert, he paid a Bank guarantee of 500 euros a month for two years, and, adding the attorney expenses, he spent more than 20.000 euros due to the fine.[106] At the same time, the *Junta* approved a long-term project directed by Palmqvist and Martínez-Navarro.[107] That summer (2005), several local and national newspapers devoted articles to the excavations carried out by their team.[108]

Little by little, year by year, Gibert lost control over research in Orce. He had been denied permits since 1987 but up until 1995 he controlled the only team that could perform large palaeontological excavations in the Orce area and specifically in Venta Micena and until 1997 he also controlled the VM-0 bone itself. Jordi Agustí and his group also went to Orce often but they limited themselves to working on microfauna and dating, Agustí's speciality. Yet, with the Orce remains trip back to the town and with the split in Gibert's team, a new group led mainly by Martínez-Navarro and Palmqvist, and with the support of Agustí and Toro, appeared to compete for the sites and the public discourses around it. Later, the creation of the *International Commission* positioned Gibert as one of the scientific heads of the research, together with

102 Muñoz, *Granada Hoy* 2005.
103 Trillo/Muñoz, *Granada Hoy* 2007.
104 Interview with Lluís Gibert, 2012.
105 Redacción, *Granada Hoy* 2005; Muñoz, *Granada Hoy* 2005; J. M., *Granada Hoy* 2005.
106 Interview with Lluís Gibert 2012.
107 Utrera, *El Ideal de Granada* 2005b and S.S./J.M., *Granada Hoy* 2005.
108 For instance: EFE, *ABC* 2005 or Redacción GD, *Granada Digital* 2005.

Toro and Agustí. Yet, this system, with scientists that were at odds working together, did not have much of a future. Gibert himself recognised that they had "no choice but to accept" a structure which was "unlikely to work".[109] In addition, with Pascual Rivas, a scientist that was clearly against Gibert's views, as head of the Commission, there were specific efforts to reject Gibert's proposals. The Commission and the new research team not only dismissed Gibert's proposals but even set him apart from a larger research project in which he theoretically worked. The 2003 monograph was a clear reflection of this marginalisation: it was a report from the last three years' excavations, in which Gibert had participated in one way or another, signed by the two other scientists in charge of the excavation, together with the leader of the team, who had just split from Gibert's own team. The subsequent *Junta* penalty that banned him from being head of any project was the final straw in a broader process of marginalisation.

To sum up, after the Orce conference, it seemed that Gibert had convinced almost everybody of the Orce Man's validity and the relevance of his research project, but the emergence of a new research team, and also the pushing aside of Gibert's ideas regarding the controversial fragment, led to his total isolation. As a scientist, he worked in the *Institut* and went to Orce every summer, but he was marginalised in that same *Institut* and did not have any real role in the Orce area's research project. As he stated in an interview in 2006, he had been forced to "retreat into the background" of the Orce research.[110]

∴

In 1997, two internationally noteworthy publications (the *National Geographic* and *Science News*) devoted one article each to the 'First European' issue in which Orce and Gibert's views appeared, or even formed the core of the pieces. In the *National Geographic*, the Venta Micena remains appeared as one of a series of possible candidates for the 'First European' title, with an image of the controversial fragment and other remains and with an emphasis on the controversial side of the claims made.[111] Some years later, in 2004, the *National Geographic* published a book called *The Human Story* in which the Venta Micena remains appeared again.[112] On the other hand, in the *Science News* piece,

109 "no nos quedaba más remedio que aceptar", "difícil que funcionara", Gibert 2004, 102 and 104.

110 "... retirarse a un segundo plano", Valenzuela, *ABC* 2006.

111 The article was part of a series of articles entitled 'Dawn of Humans'. Gore 1997.

112 Sloan 2004, 45.

Orce and Gibert were the main topics. The journalist cited the conference, the controversy with its three dissenters, and the new research team, as well as remarks made by some of the main characters in the Orce story such as Gibert, Turq, Roe, Carbonell, Roebroeks, or Tobias.[113] It was a detailed account of the more recent part of the controversy that showed the extent to which Orce could raise interest in international forums as well as pointing to the international impact of the conference. Several issues are revealed here. Firstly, some of these articles confirm the statements made at the conference, which show to what extent the international community was not just polite during the event but apparently followed some of Gibert's claims. Notably, quotes by Roe or Tobias are representative of this trend of believing Gibert despite maintaining serious doubts about VM-0; the controversial fragment. They also show Gibert and his team's constant desire to spread (and their success in spreading) their ideas regarding the Orce Man and the 'First European' more generally. For instance, in a letter to Gibert, Tobias himself hoped that what he said to the newspapers was helpful for fulfilling this desire.[114] The conference could thus be seen as a great success in this regard. The dominant presence of the famous and controversial fragment in these articles shows to what extent it was a central part of Gibert's arguments more than 20 years after the fragment was found.

Similarly, in Spain the few pieces that appeared during this period treating the Orce issue in passing showed how it could still be an interesting issue for journalists and the public. For instance, in August 1998, *El País* published a brief interview with Juan Luis Arsuaga, one of the co-directors of the Atapuerca research project. Despite the fact that only two out of eight questions dealt with something related to Orce, the journalist thought it appropriate to title the interview "JOSÉ (sic) LUIS ARSUAGA, ARCHAEOLOGIST: the Orce remains are not human, I have no doubt".[115] This shows how Orce was still publicly well-known in 1998, and even more appealing to readers than the Atapuerca research itself, specially when controversial statements were at stake. Again, in 1999, *La Vanguardia* interviewed Daniel Turbón, human evolution professor at the *Universitat de Barcelona*, who published an article after the conference praising Gibert's efforts and the Orce Man's validity. This time, Turbón was much more critical and in one of the two questions related to Orce (out

113 Bower 1997.

114 AJG-ICPMC: Letter from Tobias to Gibert, 18 November 1999.

115 'JOSÉ LUIS ARSUAGA ARQUEÓLOGO "Los restos de Orce no son humanos, no tengo ninguna duda" '. Román, *El País* 1998.

of seven questions in total); he stated that the Orce Man was not completely proven to be in fact a hominid.[116]

In February 2000, after the Spanish media attention received by some spectacular findings in the Georgian site of Dmanisi, Gibert sent a series of letters to several Spanish newspapers, including *ABC*, *El Ideal de Granada*, and *Diari de Sabadell*, and some journalists from different radio stations.[117] In some of these letters, Gibert presented the Orce research itself, the 'possession' of the First European, and, in his own words, "an alternative discourse" to that "presented by the Atapuerca researchers" which was "dominant and almost the only one in the media."[118] Gibert concluded by stating that behind the "scientific discourse, which we can control, there is a media one that escapes us completely and on it depends the social reach of the discoveries as well as the promotion of tourism", and asked the journalists to echo the Orce research.[119]

These letters reveal several issues. Firstly, that Gibert was deeply worried about Orce's public presence and, more generally, the palaeoanthropological discourse that appeared in the media. Both the 'war' with the Atapuerca research team and the promotion of tourism were issues that affected this discourse and, at the same time, the public scope of Gibert's own research and therefore its funding. Such was Gibert's concern with the media that in 2000 he even participated, together with his son Lluís, in a conference devoted to science and society with a presentation on the relationship between research in Orce and the media.[120] Likewise, throughout his popularisation book, which will be addressed later, Gibert also highlighted the prominent role of the media in the Orce controversy and the difficulties in generating a debate due to 'backwardness' in Spanish science.

Secondly, a very interesting aspect that these letters reveal is the decline of media reportage on Orce. After periods when there was a lot of media attention

116 Wirth, *La Vanguardia* 1999.

117 AJG-ICPMC: Letter from Gibert to Redacción ABC, 7 February 2000; Letter from Gibert to Redacción El Ideal, 7 February 2000; Letter from Gibert to Redacció Diari de Sabadell, 7 February 2000; Letter from Gibert to Andrés Aberasturi, Radio Nacional, 28 February 2000; Letter from Gibert to José Luis Cebrian, Onda Cero, 28 February 2000; Letter from Gibert to Carlos Herrera, Radio Nacional, 28 February 2000.

118 "discurso alternativo", "expuesto por los científicos de Atapuerca", "dominante y casi único en los medios de comunicación." AJG-ICPMC: Letter from Gibert to Carlos Herrera, Radio Nacional, 28 February 2000.

119 "de este discurso científico, que podemos controlar, hay otro mediático que se nos escapa totalmente y de él depende la proyección social de los descubrimientos así como la promoción turística." *Ibid.*

120 Gibert/Gibert 2000.

for the research (such as the original discovery, the subsequent controversy, or the Orce conference), during the late 1990s and early 2000s, interest from media diminished and what remained was mainly local. Newspapers from Granada, like *El Ideal*, *Granada Hoy*, and *Diario de Granada*, or the Andalusian edition of *ABC* followed the development of the dispute regarding the excavation permits and Gibert's final exclusion. These newspapers followed this dispute not only for the scientific side of it but also specially for its relations with Andalusian politics. Attracting the media in general and due to a more 'scientific' discourse was more difficult. To reverse this situation, Gibert himself had to promote the Orce research even more actively.

On 25 January 2001, Gibert gave a conference in Cartagena, the nearest city to Cueva Victoria, which was reported on by some local newspapers (see map 1).[121] During the event, Gibert presented new findings from Cueva Victoria and stated that the "First European was a Carthaginian".[122] According to the local newspaper *El Faro de Cartagena*, "during his meeting with the media, Gibert explained that in order to excavate in Cueva Victoria he had a budget of 500,000 pesetas [around 3,000 euros]."[123] Gibert then said that he could "not request more" but that "with a million pesetas more, I could go to Africa and compare the remains found in Cartagena with those from the same period in Africa."[124] Gibert finally praised the Town Council for the support received so far but stated that with a little more money, conditions could improve greatly. Once again, the 'oldest' discourse was accompanied with the 'lack of funds' discourse.

The next day, 26 January, Gibert had a meeting with José Cabezos, Cartagena's heritage minister. After the meeting, the politician again talked to the media to announce that the Town Council would give 2 million pesetas to the Cueva Victoria project for the following year.[125] Once again the 'look what we've found' strategy worked well for Gibert. Moreover, this turn to Cueva Victoria, with remarkable 'local' attention for the research, shows the extent to which Gibert and his team were banned from Orce and had more time and effort to dedicate to Cueva Victoria, which had always been one of Gibert's

121 S.A., *El Faro de Cartagena* 2001.

122 "El primer europeo fue cartagenero", Elgarresta, *La Verdad Digital* 2001a.

123 "Durante su encuentro con los medios de comunicación Gibert explicó que para excavar en Cueva Victoria dispone de un presupuesto de 500.000 pesetas." S. A., *El Faro de Cartagena* 2001.

124 "no puedo exigir más", "con un millón de pesetas más podría ir hasta África y comparar los restos encontrados en Cartagena con los autóctonos de la misma época", S. A., *El Faro de Cartagena* 2001.

125 Elgarresta, *La Verdad* 2001b.

research projects but with much less media attention. In addition, for many years, since the early problems after the phalanx found in 1985, Gibert and his team never had any problems with excavation permits in Cueva Victoria.[126] As one journalist, who criticized the permits refusals to Gibert, put it sometime later, due to the conditions in Orce, Gibert took "his pick and his shovel elsewhere. In this case to Murcia", where Cueva Victoria was located (see map 1).[127]

Yet, as late as 2005, Gibert and his son Lluís published quite a detailed report in the magazine *Current World Archaeology*.[128] In this report, entitled 'First People in Spain', the Giberts presented the famous Orce Man, the stone tools, and the other supposed hominid remains found in the Orce area. They also developed a 'new' model of the colonisation of Europe through Gibraltar, and also supported all of this evidence with their Cueva Victoria findings.[129] While Gibert had become more and more isolated from the excavation projects in the Orce area and the media's interest in Orce had declined, he again returned to the public arena using the Cueva Victoria finding as one of his main arguments.

4.4 The Hominids That Came from the South: Gibert's Popular Science Book

In 2004, Gibert published a popular science book on the Orce Man entitled *The Orce Man: The hominids that came from the south*, which was presented in Granada.[130] This was the third book on Orce addressed to the general public after the small one published by Bienvenido Martínez-Navarro in 1993, and the 2002 book by Domènec Campillo, which focused more on the anatomical characteristics of the bone but also reviewed the story of the Orce Man.[131] As previously mentioned, Gibert had also signed contracts to write two other popular science books that were never published. Both failed attempts occurred during peaks in Gibert's and Orce's popularity: just a few months after the announcement of the discovery of the Orce Man and shortly after the Orce conference.[132] In both cases, subsequent public controversies probably

126 Personal Communication Lluís Gibert.
127 "con su pico y su pala a otra parte. En este caso a Murcia." Rodríguez, *La Opinión de Granada* 2007.
128 Gibert/Gibert 2005.
129 Interview with Lluís Gibert 2012.
130 Soria, *ABC* 2004.
131 Martínez-Navarro 1993; Campillo 2002.
132 For more detail on the press attention to the controversy see Annex 2.

prevented the publication of the books, either because Gibert had to work on his defence and recovery or because the editors did not wanted to continue with the contract.

Finally, this time, Gibert managed to publish a popularisation book. Explaining his story at length, Gibert's book was divided into four parts. The first part started with a chapter devoted to the story of the discovery of VM-0 (Chapter 1 of Part 1: 1/1), its public presentation (2/1), the start of the controversy after the visit to Paris (3/1), and the development of the controversy (4/1 to 6/1), up until the international conference and the split in his team (7/1). Afterwards, Gibert detailed the Orce remains (8/1) and the age of the sites (9/1 and 10/1), paying special attention to Venta Micena (11/1) and Cueva Victoria (12/1). Gibert then devoted chapter 13 of this first part of the book to the life of the hominids in Orce and 14 and 15 to hominid arrival to Europe. The second part of the book was entirely devoted to reviewing the 'big picture' of human evolution (1/2), from the Pliocene (2/2) to the genus *Homo*, which included a small section on Orce (3/2), and the rest of early hominid phylogeny (4/2). The third part dealt with the anatomical characteristics of the controversial cranial fragment VM-0, the different hypotheses presented since then, the debate around it, and a comparison with some other controversies in the history of human palaeontology (chapters 1, 2, and 3 of part 3). Finally, a shorter forth part was devoted to the history of Orce's *Museo de Paleontologia y Prehistoria*. Apparently, the book had two prefaces by Phillip Tobias (figure 13) and Yves Coppens (figure 11) but they were not included due to a mistake by the publisher. It is also worth noting that throughout his book, Gibert cited the Spanish versions of other well-known popular science books like Roger Lewin's *Bones of Contention*, or Donald Johanson's *Lucy: The Beginnings of Humankind*.[133]

Gibert's book linked his story of the Orce Man controversy to the scientific study of the bone and the sites themselves, the claims of hominid arrival to Europe, the review of other controversies in the history of palaeoanthropology, and the 'big picture' of human evolution. This was done in a very mixed and even disorderly way that gives us clues about the aims and strategies Gibert had when writing the book and at the same time shows a particular moment in Gibert's research in 2004.[134] Like many other palaeoanthropologists, in his book Gibert mixed his own story with his scientific claims in a clear move to

133 Personal communication with Lluís Gibert 2014 and Gibert 2004, 445.

134 Regarding this disorder, it is worth noting that together with the omission of Tobias and Coppens' prefaces, the book was published without final corrections by Gibert. Personal communication with Lluís Gibert 2014.

legitimate his position in both.[135] The history of the discovery and the controversy of course is told in a dramatic way in which Gibert exposes the negative intentions of his 'enemies', his "paradigm adversaries" (de Lumley, Agustí, the three professors, Rivas, the Atapuerca team, Moyà-Solà, Martínez-Navarro, Palmqvist ...), the injustices he went through, and his difficulties in getting funding and excavation permits. Gibert presented a narrative of 'suffering' and of 'fighting' against his opponents in order to legitimate his research. Sentences like "somebody wanted to sink Orce's future" are typical in Gibert's book.[136] Like many other researchers that are positioned or that position themselves as 'revolutionary' or even 'alternative' or 'fringe' scientists, Gibert established and presented a narrative of villains that were against his research and persona.[137] This presentation helped Gibert to define the Orce Man story as an unfair and unequal cause, and therefore a cause worth fighting for. This highlights the necessity to study not only how a discourse presents itself, but also how presents its opposition.[138]

This narrative was combined with descriptions of the anatomical characteristics of the Orce remains and the dating of the sites. This kind of mixture in a popularisation book allowed him to present 'his' story of the Orce controversy together with 'his' controversial scientific claims, making sure that both interpretations were those placed in the public domain. This position was very clear in the book but, throughout the controversy, and in different media, Gibert reviewed time and again the way that the site and the bone were discovered, how the controversy started, and how he had worked ever since to demonstrate that he was right.[139] As C. S. McConnell argues in his PhD thesis on the Big Bang-steady state controversy, the narratives created by the actors in such controversies help to shape the way that knowledge about them is created.[140] Another good example is "the inclusion of explicitly historical accounts" in the

135 There is some historical analysis of the uses of these popular science books in palaeoanthropology: especially Hochadel 2013c, but also Hochadel 2013b, 211–252, and Hochadel 2009. There are also hundreds of popular science books written by palaeoanthropologists that very often mix the history of excavations and discoveries with purely scientific claims, almost since the very beginning of the discipline itself. Some examples include: Leakey/Lewin 1977; Johanson/Maitland 1982; Pitts/Roberts 1997; Lumley 1998; Arsuaga/Martínez 1998; and Coppens 2005.

136 "adversarios de paradigma", Gibert 2004, 94. "Alguien quería hundir el futuro de Orce", Gibert 2004, 95.

137 A similar example: Pruitt 2011, 205–6. An interesting approach to 'alternative' archaeology with case studies and firsthand experience can be found in Simandiraki-Grimshaw 2012.

138 Lyne/Howe 1986, 144.

139 For instance: Gibert, *La Vanguardia* 1993; Gibert 1999a; and Iglesias/Gibert 2003.

140 McConnell 2000, 253.

participants' scientific papers in the early development of the theory of relativity in Germany, which, according to historian Richard Staley, "argue strongly for the centrality and necessity of such discourse for the development of physics".[141] Yet, in the Orce case, Gibert somehow tried to use the history of the controversy to shape knowledge that was mostly ignored by the scientific community. Like many others, the history of the Orce Man was a 'tool' used in the Orce Man controversy, helping Gibert to reinforce his position before his rivals and to get more supporters for his cause.

Gibert also cited other controversies in the history of palaeoanthropology as part of his discourse. In his book, he reviewed the Altamira rock paintings dispute and the Taung Boy discovery and presented them as examples that show the "weight of authority in science". According to Gibert, as happened in the Orce case, the Altamira problem was a problem of authority. For Gibert, the expert send to examine the paintings, Edouard Harlé, was not a specialist, and despite of that, his verdict influenced the French authorities that were against the validity of the paintings. In Gibert's view, and has happened in his own case, the French scientists gave more credibility to Harlé's statements than to Sautuola and Vilanova's claims, without bothering to visit the cave themselves. Something similar happened, for Gibert, in the Taung story. According to Gibert, the Taung Boy story is very similar to Orce's since both discoveries were 'premature' and had critics among the scientific authorities, 'schemes' involving disciples, resistance, and finally new findings.[142]

As explained in the introduction, in his reviews of both controversies, Gibert claimed that the scientists involved (Sautuola and Dart) held the real truth since the beginning and had to fight for years against the establishment which blocked scientific progress because of different interests. For Gibert, both controversies ended due to statements from scientific authorities (Cartailhac's rectification and Le Gros Clark's report) and due to new discoveries (more cave art in France and more fossils from South African caves). This led Gibert to state that in Orce they had both the new findings and the new publications, but "for the media reality" of the case, which "moves politicians and stimulates investments", finding "a new human remain" was paramount. Yet, for this to happen, it was necessary to continue the excavations in Venta Micena, which had remained almost untouched for 20 years.[143] Gibert used these well-known

141 Staley 1998, 298.

142 Gibert 2004, 417–418: "prematuro", "intrigas".

143 Gibert 2004, 418–419: "para la realidad mediática", "mueve a los políticos y estimula inversiones", "un nuevo resto humano".

controversies to present his case and to request more excavation permits and financial support.

This was not the only use of these stories. Since very early in the Orce Man controversy, Gibert used the concept of 'paradigm' and 'paradigm change' as just another rhetorical tool to legitimate his interpretation. He quoted the famous book by the philosopher of science Thomas S. Kuhn, *The Structure of Scientific Revolutions*, to illustrate how the Orce Man could be considered as causing a change of paradigm.[144] The above cases served Gibert to show other instances of paradigm change that resulted in the blocking of scientific research. Yet, just a brief look at the scholarly literature on the history of science shows us that these controversies hide much more than the simple struggle between a scientific hero who holds the truth, which will change the paradigm, and the scientific establishment that blocks his/her progress. In both cases, historians have shown that not the entire so-called establishment was against the new claims, that the proponents, the paradigm changers, also had their own particular interests, and that these interests were not always related to the search for 'scientific truth' and change the paradigm. For instance, in the Altamira case, they were related to the support of religious views, and in the Taung Boy case, the aim was to legitimate peripheral scientific positions, since Dart worked in South Africa.[145] Using these controversies also legitimated Gibert's persistence in the validity of his claims and the validity of the cranial fragment, as if they formed another 'change of paradigm'. Accepting this, it was only a matter of time before they would be generally accepted and shared among the community, thus justifying his scientific persistence in relation to the Orce Man's famous remains and also explaining his rivals' opposition to his claims. Within this wider 'change of paradigm' frame Gibert found easy answers for the questions of why the Orce Man was not accepted by the scientific community and why he was still insisting on its validity. Again, presented that way, the Orce Man appeared as a worth cause to fight for.

The use of Kuhn in palaeoanthropology was in fact quite common during the 1970s, 1980s and 1990s, especially in relation with the controversies over the supposed 'paradigms' of the First European and the First American. Not only Gibert, but several other scientists had used Kuhn's philosophy in the justification of their positions and 'their' discoveries. As Oliver Hochadel argues, Kuhn is very useful for these palaeoanthropologists because helps to 'solve' problems

144 Gibert 2004, 415–419. Reference to Kuhn: Gibert 2004, 56 and J. A., *Diari de Sabadell* 1986c. Fahnestock 1989, describes a similar situation in the Clovis vs. Pre-Clovis controversy.

145 Richmond 2009 and Moro Abadia/Pelayo 2010.

over the interpretation of remains and over these wider controversies. As Gibert's example illustrates, the use of paradigms also helps to legitimate positions in long controversies.[146]

In addition, the use of these well-known controversies side by side with the Orce Man controversy also placed Gibert's claims together with top research in the history of palaeoanthropology. When Gibert talked about Eugène Dubois or Raymond Dart or, at the Spanish level, Don Marcelino Sanz de Sautuola, he placed himself among some of the great names of science, highlighting the significance of his research and claims. Somehow, being part of History with a capital 'H' legitimated Gibert sticking with his research in Orce. As he admitted, "Raymond Dart's story is very familiar to me".[147] Just before his death, Gibert published a children book titled 'Do you want to be palaeontologist?', which repeated, for a complete different audience, the main features of Gibert discourse of marginalization and the role models of Altamira and Dart.[148]

In a press conference, Gibert even said that what happened with Orce was the same thing that "happened to Galileo, that there will always be someone who will not believe [the claims made]" no matter how much evidence is gathered. The local Granada newspaper *La Opinión de Granada* used a suggestive title for its report on the press conference, which was illustrated with a caricature of Gibert: 'The Orce Galileo' (figure 17).[149]

4.5 The End of a Long Controversy

In 2004 and 2005, the newspaper *El Mundo* published two online interviews with questions from readers for Juan Luis Arsuaga and José Maria Bermúdez de Castro, both co-directors of the Atapuerca research project.[150] Despite the fact that in both cases a question was devoted to the Orce controversy, there was no more significant attention to the Orce Man and no more headlines. The Atapuerca researchers both stated that the famous Orce remains were not human but animal. Bermúdez de Castro highlighted the value of the Orce area sites despite the controversy and Arsuaga assumed that hominid remains

146 Hochadel 2018.

147 "La historia de Raymond Dart me es muy familiar", Gibert 2004, 417.

148 Gibert 2007.

149 "...que le pasó a Galileo, que siempre habrá alguien que seguirá sin creer" and 'El Galileo de Orce' Rodríguez, *La Opinión de Granada* 2007.

150 Redacción, *El Mundo* 2004 and Redacción, *El Mundo* 2005.

El Galileo de Orce

El científico catalán alcanzó un hallazgo histórico; fue puesto en duda por el gremio y enviado al 'exilio'. Ahora, un nuevo descubrimiento le da la razón

Un día de 1976, José Gibert se plantó en un remoto, por entonces, pueblo de la provincia de Granada. Viajaba en un Land Rover y no venía a ciegas sino que, como paleontólogo experto tenía la corazonada, y probablemente algo más, de que aquí, en algún sitio del sur de la península Ibérica, había tela que cortar. En su profesión, eso significa que se podían encontrar restos dignos de estudio y, sobre todo, respuestas a más de un dilema antropológico que trae por la calle de la amargura a miles de estudiosos. "Un geólogo entrenado como yo sabía que había mucho por descubrir en esta zona", dijo.

Nunca antes nadie se había fijado en Orce. Nunca antes de que Gibert se sentara frente a esa enorme hondonada en Venta Micena (pedanía de la localidad), contemplara la línea del horizonte y tuviera un pálpito: ése que colocó al pequeño y olvidado municipio granadino en la primera división de los yacimientos arqueológicos del mundo.

Gibert pensó que aquella superficie había sido hace un millón largo de años un lago donde habían aliviado su sed los animales de la época y por extensión, donde se habían arrimado los hombres, que por entonces ya caminaban erguidos, para cazar, beber o cualquier otra actividad. El hiperactivo profesor catalán llegó a la conclusión de que había topado con un kilométrico y antiquísimo cementerio, que debe ser para un paleontólogo lo que Las Vegas para un ludópata entregado.

La comunidad científica, sus colegas, no suelen tolerar a los que hacen ruido y nunca le vieron con buenos ojos

Así que con poco más de treinta años, un joven Gibert, buscó financiación y se puso a excavar con determinación. No tardó demasiado en encontrar algo, no, en términos científicos y sobre todo, teniendo en cuenta la importancia que tuvo el hallazgo. En 1982, seis años después, tropezó con el 'VM0' (Venta Micena Cero). Algo que en teoría no debería estar allí, un trozo de cráneo. En eso siempre estuvieron todos de acuerdo.

Convencido y entusiasmado, el profesor, que ya por entonces era un especialista en el Cuaternario, puso todo su empeño en reforzar su tesis. Montó una excavación de grandes dimensiones en la zona, se compró una casa y se empadronó. Orce le enamoró en todos los sentidos.

El pueblo le correspondió. Le hicieron hijo predilecto e incluso le pusieron su nombre a un museo. Todo iba sobre sus ruedas pe[…] sus colegas. El gremio, lo que se llama la comunidad científica, no suele tolerar a los que hacen ruido; nunca le vio con buenos ojos.

Los enemigos le nacieron, y muy poderosos. Su hallazgo, que se completó con toda clase de huesos e instrumentos, no se quedaba en el cráneo sino que tenía implicaciones que afectaban a otros. Gibert estaba convencido de que, si las cifras eran ciertas y el cráneo tenía justo 1.300.000 años, el hombre entró en Europa desde África a través de la península ibérica, y no por Eurasia, tal y como siempre se ha creído. Esa teoría le enfrentaba directamente a Atapuerca, otro de los yacimientos potentes, que competía por ser el más antiguo de España.

En 1995, Gibert convocó el Congreso Internacional de Paleontología en Orce. El orgulloso anfitrión estaba dispuesto a que se evaporara cualquier duda sobre su descubrimiento. Organizó un acontecimiento de campanillas, lo nunca visto en Orce. Carpas, coches y 300 científicos de 18 países; entre ellos los más insignes de la tierra en eso de estudiar fósiles. Esa cita pareció darle la razón. Los paleontólogos, como ente, se inclinaron por creerle, pero como científicos siguieron despellejándole. Empeñados, como estaban, en rebajar el cráneo de categoría y […] pecie de borrico prehistórico.

De manera que el vivo Gibert se encontró de nuevo prácticamente sólo frente al inmovilismo de la élite científica. "Poco riguroso" era el calificativo más suave con el que le obsequiaban; "charlatán", el más incómodo; había, y hay, cierto interés porque la sociedad percibiera algo así como un locuaz vendedor que quiere colo[…]do a una tarima.

A esa situación se unió entonces la aparición de la Junta de Andalucía, que a finales de los 90 se animó a ponerle zancadillas. Gibert recibió entonces con los brazos abiertos el papel de víctima, algo que otros muchos no le han perdonado. "Me quieren echar de este pueblo", dijo en alguna ocasión. Algo de razón no le faltaba.

Al fin y al cabo, la Junta le impidió volver a excavar, le impuso una multa de medio millón de euros (aunque luego se la rebajaron), cambió su museo de nombre y le entregó la dirección del yacimiento a otro (Isidro Toro). Después llegaron encontronazos de toda clase. Como cuando le impidieron que 'importara' estudiosos americanos a la excavación. Total que, con todo el dolor de su corazón, acabó yéndose con su pico y su pala a otra parte. En este caso a Murcia.

José Gibert nunca fue un científico al uso. Demasiadas cosas le diferenciaban de sus colegas. Paseaba cierta dosis de extravagancia, hablaba demasiado y, sobre todo, había tenido éxito. "Le sonó la flauta", prefiere decir alguno. En cualquier caso, el tiempo le ha dado la razón.

Hace unos días, ofreció una rueda de prensa para presentar su nuevo descubrimiento: habían encontrado en Tarragona el cráneo de una niña muy similar al encontrado en Orce que demostraba que siempre había estado en lo cierto y que el hombre dio el salto y se plantó en Europa a través de la península ibérica.

"Así se acaba con todas las polémicas, pero como le ocurrió a Galileo, habrá gente que seguirá sin creérselo", dijo el profesor. Se le veía feliz pero envejecido y con la mirada fatigada por los años y la tarea. No era la primera vez que se miraba en el espejo de Galileo, el primer científico moderno, un ejemplo de arrojo frente a la estrechez de miras.

Ahora que las publicaciones más prestigiosas se ponen de su parte y tiene todo de cara, él desea acabar su carrera en Orce, volver a Venta Micena antes de su jubilación. Regresar a aquella hondo[…]

Una larga trayectoria vinculada a Granada

La carrera de José Gibert no se puede entender sin su hallazgo en la localidad de Orce.

1976. Viaja a la provincia de Granada desde Cataluña y descubre el yacimiento de Venta Micena.

1982. Durante las excavaciones, encuentra lo que parece ser un fragmento de cráneo humano de 1,3 millones de años.

1995. El Congreso Internacional de Paleontología avala su descubrimiento.

2004. La Junta le impide seguir trabajando en Orce.

2007. Un cráneo hallado en Tarragona reafirma su tesis.

FIGURE 17 Gibert's caricature with the suggestive title 'The Orce Galileo'
FROM RODRÍGUEZ, *LA OPINIÓN DE GRANADA* 2007

would be found someday in Orce. These interviews show a clear decline in interest in the Orce Man issue, yet the fact that the questions on Orce were posed by the general public also shows that the Orce Man still captured the attention of those with palaeoanthropological interests.

In 2005, Gibert got significant backing from Catalan institutions when he was awarded the Narcís Monturiol Medal for scientific achievement.[151] Gibert's scientific isolation throughout the country, including in his own institution (the *Institut de Paleontologia de Sabadell*), and his political isolation in Andalusia were not reproduced by the Catalan government. One of the reasons for this could be the ideological closeness between Gibert and the government in Catalonia since 2003 when a left-wing coalition of three parties, the 'Tripartit', came to power.

In January 2007, Gibert and his son Lluís held a press conference in Granada to present the recent research carried out by the team. The announcement of the press conference pictured an image of the cranial fragment VM-0 and its title was 'The Orce Man: New findings, dating of the sites and the surroundings'. The press conference received great media attention at the Andalusian level and to some extent at the national level.[152] After again explaining the story of the controversy, Gibert, and later the press, highlighted two new events that apparently finally put an end to the controversy. The first was an article published by Domènec Campillo and some of his collaborators in which they presented the cranium of an ancient Roman girl found in Tarragona that had exactly the same cranial crest that triggered the controversy back in the 1980s.[153] This could be proof that the Orce Man was definitively a man. The second was a letter by Emiliano Aguirre, first director of the Atapuerca site research and now retired, who admitted that this finding was definitive proof that Gibert had been right since the beginning. Most of the newspapers presented this new evidence as confirming his theory and claimed that now the *Junta* must act in consequence and allow Gibert to work in Venta Micena. Gibert himself stated that now "the controversy over this issue [the cranial fragment] is over" and that "the truth has prevailed".[154]

Yet, that summer, Gibert again was not granted an excavation permit and the penalty imposed on him was still being appealed in the Supreme Court of Andalusia.[155] By the end of September 2007, while excavating in Cueva Victoria,

---

151 Utrera, *El Ideal de Granada* 2005a.

152 'El Hombre de Orce: Nuevos hallazgos, datación de los yacimientos y su entorno' AJG-ICPMC: Triptic Press Conference, 15 January 2007. For the attention see: EFE, *El País* 2007; Romero, *Granada Hoy* 2007; Cambril, *La Opinión de Granada* 2007; Rodríguez, *La Opinión de Granada* 2007; Ladrón de Guevara, *El Ideal de Granada* 2007; Sáiz-Pardo, *El Ideal de Granada* 2007; Gómez, J. E. *El Ideal de Granada* 2007.

153 Campillo et al. 2006.

154 See Cambril, *La Opinión de Granada* 2007 and "La controversia sobre este tema se ha acabado", "La verdad se ha impuesto", Ladrón de Guevara, *El Ideal de Granada* 2007.

155 Cappa, *Granada Hoy* 2007.

he felt sick and went back to Barcelona sooner than expected. A week later, on 7 October, Gibert died of lymphatic cancer. His ashes were spread in Venta Micena.[156] Several obituaries were dedicated to him in numerous newspapers. Even Salvador Moyà-Solà signed the *El País* obituary, where he stated that despite the scientific confrontation between them, Gibert showed great enthusiasm, vocation, and optimism throughout his career. According to Moyà-Solà, Gibert was crucial to the recovery of momentum that the *Institut de Paleontologia* experienced with the Orce findings and after Crusafont's death.[157]

After Gibert's death, the whole controversy around the Orce remains changed. He was the primary and most famous defender of the so-called Orce Man, the famous cranial fragment that from then on was more or less forgotten, despite some popular articles signed by some of Gibert's collaborators such as Domènec Campillo.[158] His son Lluís tried to continue his work in the Orce area and presented excavation projects for Venta Micena that so far have been refused. In any case, Lluís's scientific work focuses more on geology and dating methods than on the hominid presence in Orce.[159] In short, with Josep Gibert i Clols's death, not only a palaeontologist's life came to an end but also a very long-lasting controversy.

156 Sanz, *ABC* 2007; Moyà-Solà, *El País* 2007; Planas, *La Vanguardia* 2007; Europa Press *VilaWeb* 2007; EFE, *El Mundo* 2007. Several palaeoanthropologists, like Gibert, have spread their ashes in 'their' sites or want to do it in the future. Hochadel 2013b, 191.

157 Moyà-Solà, *El País* 2007.

158 Campillo 2010, see also Cardeñosa 2007.

159 Personal Communication with Lluís Gibert 2014.

CHAPTER 5

A 'First' American to Compare: The Pedra Furada Controversy

After reading my account of the Orce Man story, some might say, and some had said, that the Orce Man is a very special, unique story, a rarity in scientific practice, and even in palaeoanthropology. What these opinions usually argue is that the way that the discovery was presented, the way the controversy broke-out in the *El País* front page and the way Gibert defended his views, were not 'normal', and were consequence of a very particular setting. Though I could agree that the Orce Man story was deeply shaped by its historical context, along the book I have used many examples that show that it is not as unique at it may seem, yet I have not developed in some detail any of them. In the next lines, I will introduce a quite unknown and very similar story to that of the Orce Man in order to show better that the Orce Man is not an isolated case but even a quite common story, specially in these palaeo-sciences, that show us a lot about the way these sciences worked in the 20th century.

Instead of a possible 'First European', this is the story of a possible 'First American', found at the Pedra Furada site in Brazil. The analysis of this case is preliminary and it is much more complex than what will be presented here. In 1978, the excavations at the Pedra Furada rock shelter began. This site is located in the Serra da Capivara National Park in northeastern Brazil, in the Piauí region, very isolated and one of the poorest in the country. The excavations were led by the French-Brazilian archaeologist Niède Guidon. After growing up in Brazil, she studied archaeology in Paris. In the Serra da Capivara, Guidon and her team classified hundreds of sites with remains and rock art.[1] The Pedra Furada rock shelter was one of the biggest in the area. This research was mainly funded by the French *Centre National de la Recherche Scientifique* (CNRS), where Guidon later obtained a position.[2] In the early 1980s, some articles on Pedra Furada's rock art and dating were published in France.[3] These were accompanied by a notable

1 A more detailed account of Guidon's story in Carandell 2016 or, written from Guidon's own prespective, Drévillon 2011. More information on Fumdham's web page: http://www.fumdham.org.br/.

2 Drévillon 2011, first chapter.

3 For instance: Guidon 1983 and Guidon 1984.

 | DOI:10.1163/9789004431508_007

outreach effort, mainly in Brazil, including long magazine reports and museum exhibitions.[4]

This early popularisation effort was linked to two interesting issues. On the one hand, the promotion of the Serra da Capivara as a tourist destination, despite being very isolated, in order to improve the living conditions of the locals.[5] On the other hand, it was also linked to a need for funding discourse, mainly for protection of the archaeological sites of the Serra da Capivara.[6]

Everything changed in June 1986. That month, the British journal *Nature* published an article by Guidon and a French collaborator, Georgette Delibrias, on her findings at Pedra Furada.[7] In this article, Guidon presented evidence, mainly "abundant lithic industry", that pushed back the arrival of the earliest inhabitants of South America to 32,000 years ago, twice as old as the accepted presence of modern humans in the Americas.[8] The cover of that issue of *Nature* was devoted to Pedra Furada's rock paintings with the title 'Early Man in South America'. Together with other findings, the publication of this article reopened the long-lived Clovis vs. Pre-Clovis controversy in the late 1980s.[9] In the early 20th century, North American archaeologists claimed that the earliest human presence on the continent was the so-called 'Clovis culture', named after a site in New Mexico that was no older than 12,000 years. Yet, from the moment this theory was formulated, other scientists claimed to have found older sites. Since then, the earliest population of the Americas has been a matter of debate between Clovis and pre-Clovis proponents.[10] The Pedra Furada findings were yet another challenge to the Clovis theory, a challenge trumpeted on the front cover of *Nature*. Some scholars strongly opposed Guidon's claims while others supported them, starting an active debate on the peopling of America for some years mainly on the pages of the archaeological journal *Antiquity* and its American version *American Antiquity*.[11]

---

4 For instance: Leitao 1978; Redacción, *Folha do São Paolo* 1978; Redacción, *Folha do São Paolo* 1979; Redacción Veja 1983; Teresina (Correspondente) *Folha de Sao Paolo* 1978; Belik, *Folha de São Paolo* 1983; Belik, *Folha de São Paolo* 1984. See also Carandell 2016.

5 Drévillon 2011, 59, 136.

6 *Ibid.*, 155 and Natale, *Folha de São Paolo* 1984.

7 Guidon/Delibrias 1986.

8 *Ibid.*, 769.

9 Whitley/Dorn 1993, 626.

10 See for instance: Whitley/Dorn 1993; Fahnestock 1989.

11 Opposition: Turner 1986; Meltzer 1989; Lynch 1990; Lynch 1991; Meltzer/Adovasio/Dillehay 1994; Borrero 1995; Prous/Fogaça 1999. Support: Bray 1986; Gruhn 1987; Bray 1988; Bednarik 1989; Gruhn/Bryan 1991. See also Drévillon 2011, 89. See also Carandell 2016.

In the Brazilian public sphere, where this finding was nicknamed 'Homem do Piauí', Guidon linked the discovery to a discourse of revolution in science, of changing the traditional theory of the humans arrival to America, the Clovis theory.[12] Some months after the article was published, Guidon and her team founded the *Fundação Museu do Homem Americano*, a foundation to promote scientific research and to improve local living conditions, and started to build a museum in the area.[13] The Serra da Capivara was first recognised as a Brazilian National Park and later as a *UNESCO* Heritage Site, despite the fact that Guidon's claims were (and still are) highly disputed.

The early popularisation effort made in Brazil facilitated the widespread reception of the *Nature* article, transforming Pedra Furada into the starting point of Brazilian history, for instance, in school text books.[14] At the same time, this article was widely covered in magazines and newspapers in Brazil but also around the world.[15] The Pedra Furada Man was not only the 'First American' but also the 'First Brazilian'. Despite this, some Brazilian or Brazil-related scholars also criticised Guidon's conclusions in Brazilian newspapers and scientific articles.[16]

In 1993, Guidon and her team organised a major conference in the Serra da Capivara with prominent American scientists engaged in the Clovis debate among the attendees.[17] The conference gathered the international scientists in the isolated region of the Serra da Capivara, with site visits to Pedra Furada included. It was widely reported in Brazilian newspapers and also internationally (figure 18).[18] After the conference, three of the American attendees, that apparently did not criticized Guidon's research during the conference, published an article in the journal *Antiquity*, a well-known archaeological journal.[19] This article accused Guidon and her research team of having made several mistakes

12 For instance: Redacción *Folha do Sao Paolo* 1993. See also: Fahnestock 1989.

13 Drévillon 2011, 155 and Da Reportagem Local, *Folha do São Paolo* 1986.

14 Funari 2000.

15 In Brasil: Redacción Veja 1986; Wilford, *Journal do Brasil* 1986; Das Agéncias Internacionals, *Folha do São Paolo* 1986; Reis, *Folha do São Paolo* 1986; Redacción Veja 1987. Internationally: Hammond, *The Times* 1986; Wilford, *The New York Times* 1986; Rebeyrol, *Le Monde* 1986; De Torres, *La Vanguardia* 1986. See also Carandell 2016.

16 In the press: Bonalume, *Folha de Sao Paolo* 1986; Da Reportagem Local, *Folha do São Paolo* 1993; Neves, *Folha do São Paolo* 1995. For scientific articles: Prous/Fogaça 1999. See also Drévillon 2011, 89.

17 For a closer analysis of this Conference and its comparision with Orce's see Carandell 2015a.

18 For instance: Gurovitz, *Folha do Sao Paolo* 1993; Redacción, *Folha do Sao Paolo* 1993; Da Reportagem Local, *Folha do São Paolo* 1993; Lemonick 1993; Vincent, *The Globe and Mail* 1994; Hammond, *The Times* 1995; Wilford, *The New York Times* 1995. See also Carandell 2016.

19 Meltzer/Adovasio/Dillehay 1994.

FOLHA DE S. PAULO ciência 11 de abril de 1993 6-13

Homem do Piauí divide arqueólogos

Tese de cientista franco-brasileira de que América foi habitada há 50 mil anos ganha força; veja abaixo a polêmica

Francês não acredita no 'homem do Piauí'

OUTRO LADO

"É preciso ir lá para ver"

GLOSSÁRIO

Escavações em um dos sítios arqueológicos de Pedra Furada, no interior do Piauí

Teoria tradicional é falha, diz Guidon

OUTRO LADO

'Sítios antigos são duvidosos'

FIGURE 18 *Folha do Sao Paolo* reporting on the conference
FROM DA REPORTAGEM LOCAL, *FOLHA DO SÃO PAOLO*

during the Pedra Furada excavation and in the interpretation of the findings. Guidon and her team responded some time later in the same journal.[20]

In the late 1990s, Pedra Furada's claim lost prominence in Brazilian media, despite a permanent popularisation effort. As Guidon recognised, she preferred to communicate in a way that everybody could understand rather than only to the close circle of American archaeologists.[21] In 2013, a documentary and a book that repeated all these discursive resources were released on

20 Guidon et al. 1996.

21 Drévillon 2011, 90.

Guidon and her claims in France.[22] Recently, French archaeologist Eric Böeda, Spanish archaeologist Ignacio Clemente-Conte, and other researchers joined Guidon with new research and findings in Pedra Furada and nearby sites.[23] In an article in *Antiquity*, Guidon and these scientists claimed that new evidence of human-made stone tools pointed to occupation of the region being "more than 20,000–25.000 years ago".[24] This led the authors to the conclusion that the "currently accepted narrative of human settlement in South America will have to be re-thought".[25] Again, this new research had a certain presence in popular media in Brazil and also in Spain.[26]

The Pedra Furada case interests us for its striking parallels with the Orce Man dispute. They involved controversial claims of having evidence of the 'First' inhabitants of America and Europe and in a very similar period. Both claims were personified in scientists with strong personalities whom kept the controversies (which started with a front cover in *Nature* and a front page in *El País*) going for almost 30 years. Both lead scientists launched powerful discourses of being 'revolutionaries' changing the established 'paradigm', challenging the scientific establishment, and fighting foreign 'scientific colonisation' and, hence, as being often ignored and marginalised by colleagues at home and abroad. In addition, both controversies involved scientists that had direct contact with local and national politicians, transforming them from solely scientific into very political controversies. Finally, both cases, before becoming proper controversies, developed these discourses not only in the scientific media but very actively, and crucially, in the popular media. Both cases have also been reshaped through new research in the same area, which has again stimulated debate and popularisation, despite the fact that Gibert was excluded, and Guidon was included in this research.

Several questions dealing with the Orce case throughout this book could be raised through this comparison: how did these different specific geographical, political, or social contexts influence the way these scientific discoveries, popularisation efforts, and controversies developed? Did these specific contexts play a role in the way the researchers presented their claims to scientists and to the public? Which elements of the strategies, 'tools', and discourses used

---

22 *Ibid.*

23 Böeda et al. 2014b and Böeda et al. 2014a.

24 *Ibid.*, 927.

25 *Ibid.*, 927.

26 For instance: Redacción National Geographic España 2013; Bedinelli *El País* 2014; Aranda, *El Mundo* 2014; Romero, *The New York Times* 2014; Sotos/Madridejos, *El Periódico* 2015; Lopes, *Folha de São Paolo* 2015.

follow a pattern in both controversies regardless of their context? Which do not follow a pattern, and why? How did the gender difference between Gibert and Guidon affect, for instance, differences in 'rival' attacks?

More research is needed to answer these and several other questions. Yet, the Pedra Furada example shows how the study of some specific issues treated in this book could be successfully applied to other cases in which discovery and controversy are involved. The Pedra Furada story give more value to the Orce Man controversy as, together with other examples of similar controversies, they eliminate its uniqueness, the argument of its supposed 'not normality'. The similarity in the decisions and actions taken by Gibert, Guidon and the other actors of both stories show how their procedures were not only usual, but even 'normal' in scientific practice. It is maybe the outcome of some of their actions, or the reaction of their colleagues, shape by the historical, scientific and political contexts, what transformed these usual actions into 'unique' controversies. It is in this way, considering that the way that scientists in this controversies acted were normal, that the Orce Man controversy not only show us more about how scientific controversies work, but also shows us more about how general scientific practice worked in this historical context.

CHAPTER 6

Coda: The 'Orce Boy'

On 14 May 1984, just two days after the controversial *El País* front cover, which initiated the Orce Man dispute, a boy called Mark Zuckerberg was born in White Plains, New York. In the thirty-six years since then, the North American entrepreneur helped change the world by inventing Facebook, and he became a billionaire in the process.[1] Meanwhile, the controversial bone fragment had been almost forgotten about, except, perhaps, in the small town of Orce, where still is the museum's 'star'.[2] But in May 2013, the social change that the figure of young Zuckerberg so well represents became relevant in the most recent (though but probably not the last) 'episode' of the Orce Man story. A new discovery was presented to the public (figures 19 and 20). This time, the finding seemed unquestionable, but in the end, it too sparked dispute. More local, more virtual, less widely followed, and not as long lasting as the previous one, this new controversy involved some of the actors that have already appeared in the story, but also some new ones. If back in the 1980s scientists communicated and debated through traditional media, in 2013, Facebook, Twitter, blogs, and similar social media were the chosen channels for the unfolding of the scientific dispute. Old disputes but new media will present, at the same time, some relevant continuities and parallels, but also important shifts in human origins research, and its communication, from the previous controversy. The Orce Boy will be useful not only to follow until recently the Orce story, but also to propose some initial clues on how research worked in the early 2000s. This section aims to provide insights that could help answer some crucial questions about digital scientific communication: How, at the beginning of the 21st century, was a new discovery presented to society? Have new communication channels affected the way that scientific knowledge is presented to the public? How are scientists involved in this communication? Are 'digital' controversies different from 'traditional' ones? How much have things changed since the early 1980s?

The palaeoanthropological world had also changed in the intervening 30 years. While in the eighties everybody red Donald Johanson's book on Lucy, in the early 2000s palaeoanthropology was beginning to be communicated

1 Wikipedia Contributors 'Mark Zuckerberg'.

2 Utrera, *El Ideal de Granada* 2015.

 | DOI:10.1163/9789004431508_008

in a different way. The so-called Rising Star Expedition illustrates this quite well. In October 2013, a South African-American expedition, sponsored by Witwatersrand University and the National Geographic Society and led by the South African palaeoanthropologist Lee Berger, discovered a series of remains later attributed to *Homo naeldi* in the Rising Star Cave in South Africa. With more than 1200 hominid fossils collected, the expedition had an impressive presence in the social media, with scientists narrating the findings on Twitter and YouTube almost minute by minute. A blog, called 'Rising Star Expedition' with the subtitle 'Prehistory in the making', followed the excavations every day, featuring interviews with the researchers, reviews of the workday, and videos. Everything was shown almost in real time; everything except for the bones themselves, which were generally hidden, awaiting detailed study.[3]

∴

After Josep Gibert's death in October 2007, there were no excavations in Orce's main sites the following year.[4] From then on, a battle started between two research projects to get excavation permits for the Orce area. On the one hand, the Catalan archaeologist Robert Sala was the director of a multidisciplinary research project that included Paul Palmqvist and Bienvenido Martínez-Navarro.[5] Sala had worked with the Atapuerca team and was a very close colleague of Eudald Carbonell, with whom he wrote two popular science books.[6] On the other hand, Lluís Gibert Beotas, Josep Gibert's son, formed a team of scientists that had always supported Gibert, such as Enrique García-Olivares or palaeontologist Francesc Ribot, and international scientists like the American geochronologist Gary Scott.[7] Gibert was criticised precisely for being Gibert. Pedro Benzal, head of the *Junta*'s Department of Culture, stated that research projects in Orce "do not have surnames and are not hereditary" and that the important thing is for them to be "rigorous".[8] In response to this, the mayor of Orce, José Ramón Martínez Olivares, stated that if excavation projects are not

3 See for instance: Howley 2013a, Howley 2013b, Howley 2013c. See also: Wikipedia Contributors 'Rising Star Cave'.

4 Corbella/Sáez, *La Vanguardia* 2008.

5 Redacción, *El Ideal de Granada* 2009 or Europa Press, *El Ideal de Granada* 2010.

6 Carbonell/Sala 2000 and Carbonell/Sala 2002.

7 Interview with Lluís Gibert 2012.

8 "[N]o tienen apellido ni son hereditarios", "rigurosos", EFE, *Soitu.es* 2008. See also: Redacción, *Granada Hoy* 2009.

hereditary, then the same should go for "the *Junta*'s marginalisation and persecution of the Gibert surname."[9]

In September 2009, Lluís Gibert and Gary Scott had a publication in *Nature* on geochronology that, although based on research carried out in several sites in south-eastern Spain, had important implications for the dating of Orce's supposed hominid site and the use of hand-axes.[10] The publication was widely reported on in the Spanish traditional and online press and even had international reach.[11] The fact that Gibert's son had published in *Nature* (even if only in the section 'Letters to Nature' where the articles are brief) was highlighted in these reports. In 2011, Palmqvist and other researchers responded with a scientific article stating that they had "detected a number of errors in the interpretation of the archaeological assemblage" in Gibert and Scott's article and "discrepancies" in the fauna lists with "species that are omitted or arbitrarily changed to make the assemblage consistent with the new ages."[12] It seemed that the disagreement between sides had not ended with Josep Gibert's death.

In November 2009, excavation permits for Barranco León and Fuente Nueva 3, but not Venta Micena, were granted to Robert Sala's project for the summers of 2010 and 2011.[13] In September 2010, newspapers stated that 1.3-million-year-old arrowheads had been found in Orce, proving the presence of early hominids in the area.[14] Lluís Gibert again requested an excavation permit for Venta Micena in 2011, which was denied. In 2012, nobody excavated in Orce as permits and funding were also denied to Sala's team.[15] Finally, in January 2013, a four-year project with 330,000 euros in funding and excavation permits from the *Junta de Andalucía* for Barranco León, Fuente Nueva 3, and also Venta Micena (the site where the Orce Man was originally found) was approved for Robert Sala's team.[16] Sala highlighted in the media that Venta Micena was only important for understanding the flora and fauna of Orce before the arrival of

9 "El alcalde de Orce defiende que si los permisos de excavación no se heredan, tampoco se puede heredar la marginación y persecución de la Junta hacia el apellido Gibert.", Utrera, *El Ideal de Granada* 2009.

10 Scott/Gibert 2009.

11 Utrera, *El Ideal de Granada* 2009; Hernández, *El Mundo.es* 2009; S. S., *ABC* 2009; Madridejos, *El Periódico.com* 2009; EFE, *El País* 2009; Fountain, *New York Times* 2009a; Fountain, *New York Times* 2009b; Balter 2009; Bower, 2009; Weise, *USA TODAY* 2009; Bonalume, *Folha de Sao Paolo* 2009.

12 Jiménez-Arenas et al. 2011.

13 Redacción, *El Ideal de Granada* 2009.

14 Utrera, *El Ideal de Granada* 2010, and Valverde, *El País* 2010.

15 Utrera, *El Ideal de Granada* 2012.

16 EFE, *El Ideal de Granada* 2013.

hominids.[17] This new research team supported the idea of hominid presence in Barranco León and Fuente Nueva 3, mainly on the basis of stone tool evidence. Yet, considering Venta Micena as *only* important for flora and fauna research *before* hominid arrival reveals Sala's team's desire to dissociate themselves from Gibert's claims. The idea of there being no possible hominid presence in Venta Micena removed any chance of Gibert's Orce Man being restored.

On 1 March 2013, a press conference was organised in Granada to announce the discovery of a 1.4-million-year-old hominid milk tooth in Barranco León, one of Orce's main sites (figure 19). This tooth was presented as the first example of hominid remains ever found in Orce. Luciano Alonso, the *Junta de Andalucía*'s head of culture, and José Ramón Martínez Olivares, the mayor of Orce, chaired the event together with Isidro Toro and Bienvenido Martínez-Navarro (figure 20). Martínez-Navarro himself suggested the nickname 'Orce Boy' or 'Orce Girl' in a clear strategy to separate the finding from the old and controversial 'Orce Man'.[18] The discovery was reported on by all the major Spanish newspapers and even by some international ones.[19] Toro and Martínez-Navarro were the main co-authors of a scientific article that presented this tooth that was about to be released 'in press' five days later (5 March) in the online version of the prestigious *Journal of Human Evolution* with the title 'The oldest human fossil in Europe dated to ca. 1.4 Ma at Orce (Spain)'.[20] Jordi Agustí, José María Bermúdez de Castro (one of the Atapuerca researchers that attended the Orce conference in 1995 and an expert in fossil dentition), and Paul Palmqvist were also among the co-authors of the article. In the acknowledgements, the authors mentioned Pascual Rivas, who was thanked for his support and encouragement, and Henry de Lumley and his team, who were also thanked for their help. It is worth noting that the article was sent to the journal in October 2011 but not accepted until 21 January 2013, just six days before the four-year project with 330,000 euros in funding and excavation permits was approved for Robert Sala's team. It is not unreasonable to think that the prestigious international publication of the hominid remains helped them achieve the funding or at least legitimated it.[21] Martínez-Navarro pointed out that the article had been subject to 6 peer reviews before its acceptance.[22] The

17 Europa Press, *El Ideal de Granada* 2013b.

18 For instance: Europa Press, *El Ideal de Granada* 2013a.

19 A. G., *ABC* 2013; Cortés, *El País* 2013; Madridejos, *El Periódico* 2013; Europa Press, *El Ideal de Granada* 2013a; Europa Press, *La Vanguardia* 2013; Marot, *der Standard* 2013.

20 Toro-Moyano et al. 2013a, original version in press.

21 El Niño de Orce 2013a.

22 Cortés, *El País* 2013.

FIGURE 19 The Orce Boy tooth
SOURCE: ©JORDI MESTRE

delay between sending and acceptance, instead of a sign of difficulties and controversy, was presented and understood as demonstrating the article's value and legitimising its validity.

So, was the Orce Boy's presentation to society different to the Orce Man's? In both cases, a press conference including politicians was organised. Back in 1983, the politicians were from the *Diputació de Barcelona* and from the *Junta de Andalucía*. Now, Catalan politicians were not present but both the *Junta* and the Orce Town Council had their representatives. After a national presentation (both in the 1980s and in 2013), the discoveries were brought to Orce where they were presented to the town's inhabitants.[23] In both cases the hominid fragment was tiny but apparently unquestionable. In both cases, the remains were presented to the media, and to the scientific community, as the 'oldest' European remains ever found. In both cases, the remains were not fully or properly published in the scientific sphere at the time of public disclosure. The 1980s Orce Man was only published in the *Institut de Paleontologia*'s journal while the 2013 tooth was accepted in the prestigious *Journal of Human Evolution* but not yet published. Also, in both cases scientists received funding related to the public announcement of the discovery both before and after that announcement. In short, many supposed 'bad practices' and 'errors' supposedly made by Gibert, Agustí, and Moyà-Solà at the very begin of this story were repeated with the new discovery. The involvement of politicians, the media presentation, the supposedly exaggerated 'oldest' European claims, and the difficulty in accessing scientific information on the finding at the time of public presentation are all issues that reappeared with this new discovery. This situation again suggests

23 El Niño de Orce 2013b, and El Niño de Orce 2013c.

FIGURE 20 Presentation of the Orce Boy tooth. In the middle, Luciano Alonso, the *Junta* head of culture. Pointing at the tooth, Bienvenido Martínez-Navarro and between them, Isidro Toro
PHOTO: SERGIO ROS-MONTOYA, UNIVERSITY OF MÁLAGA

that the way the original Orce Man was presented to the public was not exceptional in palaeoanthropology or science more generally.

∴

It is worth pointing out that during the first five days after the presentation of the Orce Boy (figure 20), when most of the reports were written, the scientific article was still unpublished. Therefore, the only sources for journalists were the researchers themselves, their press releases, or their statements at the press conference or in subsequent interviews. Some information also circulated thanks to online platforms.

Apparently, the fossil tooth was discovered in 2002, when Gibert was still alive and Agustí, Martínez-Navarro, and Toro were progressively excluding him from the excavations.[24] In an article in the Spanish online science magazine

24 See for instance: Ansede, *Materia* 2013a, or Cortés, *El País* 2013.

Materia, Martínez-Navarro stated that back then, he "did not want to believe that the tooth was human" since it was found together with hundreds of other teeth from different mammals and he decided to be cautious and not to publish it.[25] In the same piece, Bermúdez de Castro explained how in 2008 Isidro Toro came to his office with the tooth in a box. As an expert in hominid teeth, Bermúdez de Castro immediately identified the tooth as "probably human".[26] After some studies, they decided to publish the finding, which finally happened in 2013.

This was the discoverers' version; a blog called *El Niño de Orce* had a different one (figure 21). It is important to take into account that this blog was written anonymously and appeared right the day after the presentation of the tooth. It was also written with a great amount of detail and a lot of information on the research performed in Orce. Generally, it was also harshly critical of the Sala, Toro, and Martínez-Navarro team and the presentation of the new discovery. On the blog, the authors defined themselves as a group of people related in one way or another to Orce and said their aim was to present all of the information on the Orce research, gathering everything that was written about it on the internet.[27] Lluís Gibert claimed that the author of the blog was a person who had been mistreated by Sala's research team.[28] Some months after the new controversy, the blog disappeared and is no longer available. According to *El Niño de Orce*, when the tooth was found in 2002 nobody labelled it or considered it human. It was later (sometime before 2008) that somebody realised it could be human.[29] While the discoverers justified this delay by saying that it was due to the detailed study to which the tooth had been subjected, critics said that had the tooth been presented earlier, Orce would have had more resources to attract tourism which would have improved the town's economic situation.[30] This flurry of comments and criticism also reached the small town of Orce. According to *El Niño de Orce*, after Martínez-Navarro presented the tooth to the locals, a lot of criticism arose about the delay in the publication.[31] The 'virtual' criticism entered the local world of Orce and reached the 'real' scientists.

During these first days after the public presentation of the tooth, there were also a lot of comments and references to the past controversy. In the press

25 "No quise creerme que el diente fuera humano", Ansede, *Materia* 2013a.
26 "Probablemente humano", Ansede, *Materia* 2013a.
27 El Niño de Orce 2013e.
28 Personal communication with Lluís Gibert 2014.
29 El Niño de Orce 2013d.
30 El Niño de Orce 2013i and Furió 2013.
31 El Niño de Orce 2013c.

FIGURE 21 *El Niño de Orce* blog header. The informed but anonymous blog vanished after the controversy

conference, Bienvenido Martínez-Navarro stated that the Orce Man was not a man. In contrast, this new Orce Boy was "undeniably" human. He concluded that in "10 or 20 years' time it will still be considered human".[32] At the same time, Gibert was remembered and even honoured on some blogs and even on radio and television.[33] Some of the comments stated that this discovery finally confirmed that Gibert was right. The day after the press conference, Lluís Gibert, Josep Gibert's son, told the press that the discredit suffered by his father in the presentation "hurt" but he was "glad" about the discovery of the tooth.[34] According to him, "in Orce there have been no ethics, and without ethics there is no progress."[35] Some days later, *Izquierda Unida*, the political party that supported Gibert during the 1995 conference, presented a petition to the *Diputación Provincial de Granada* to grant Gibert the adopted son distinction posthumously.[36] This acknowledgement was finally granted to Gibert five months later.[37]

A completely different critique was also made of the discovery. In an online economics forum, a lively discussion opened up on foot of knowledge of the 330,000 euros of public funding received by Sala's team. The main complaint was that with the economic situation in Andalusia at the time there were priorities for public funding other than archaeological or palaeontological research.[38] The title of the debate sums this complaint up perfectly: 'In Andalusia we haven't enough to eat and they go and spend the money on finding

32 "[D]entro de 10 o 20 años seguirá siendo considerado como humano", Madridejos, *El Periódico* 2013.
33 For instance: El Niño de Orce 2013f; Cuatro, Cuarto Milenio 2013; *La Otra Mirada* 2013; *Rosa de los Vientos* 2013.
34 "… me duele", "me alegra", Informativos, *Canalsur.es* 2013.
35 "En Orce no ha habido ética, y sin ética no hay progreso", Informativos, *Canalsur.es* 2013.
36 Redacción, *El Ideal de Granada* 2013a.
37 Redacción, *El Ideal de Granada* 2013b.
38 Burbuja, Foro de Economía 2013.

a tooth'.[39] The debate then went on to discuss the public or private funding of heritage research with positions on both sides of the argument. Beyond the perhaps anecdotal nature of this debate, what the critique reveals is that the economic and political situation in 2013 was completely different to that of 1983. In 2013, the euphoria of the new democracy had completely vanished. The global economic crisis and the great number of corruption cases revealed in Spanish democratic institutions (dominated by the two main parties: PSOE and PP) had led citizens to a permanent state of indignation which also affected the new Orce discovery.

When, five days after the press conference (5 March 2013), the scientific article finally appeared, it had one feature that was immediately highlighted and that caused the subsequent controversy. In a 9-page scientific article on a supposed hominid fossil found in the Orce area, the name Gibert was nowhere to be seen; there was not a single reference to his 30 years of work in the area, not even to criticise him or his theories. This is even more unusual considering that Gibert was the discoverer of Barranco León and that his team had, in 1999, published an article on a supposed hominid tooth fragment found at the same site.[40] In addition, the famous Orce Man (VM-0) was not cited in the article either, although it had been found in the same area. The strange thing is that even if they did not agree that the earlier controversial fragment belonged to a hominid, Gibert's claims of early hominid presence in the Orce area, for instance those based on stone tools, could have been beneficial for supporting their own claim that the new tooth fragment belonged to the earliest hominid in Europe. Or even presenting the previous findings as 'wrong' could have been used as a way to present the new fragment as the first 'real' Orce hominid discovery. Instead, the article completely ignored Gibert and his scientific activity in the Orce area.

Why did Agustí, Martínez-Navarro, Toro, and their team deliberately omit Gibert's name from the scientific article and, at the same time, quote it in the press conference and use it to legitimise their new discovery? Moreover, what were the consequences of this strategy? The second question will be assessed in the next section, but now let us try to explore the first one. The way that Gibert was erased from scientific research in Orce highlighted Martínez-Navarro and Toro's desire to establish a new research tradition in the area with no connection whatsoever to the old Orce Man controversy. They wanted a clean slate to reframe Orce as an area holding the remains of an *un*controversial early

39 'Andalucía, no tenemos pa' comer y se lo gastan en buscar un diente', Burbuja, Foro de Economía 2013.

40 Gibert et al. 1999d and Arribas/Palmqvist 2002.

European hominid and to present a new Orce Man (in this case an Orce Boy) that could be considered the 'First European' without being linked to the old controversial fragment. And, though it is not strictly linked to the tooth discovery, there is one more related issue worth mentioning. From 2000 to 2007, Isidro Toro was the director of the *Orce Museum*. During this time, a series of informative panels on the Orce sites were installed. Among these panels, there was one devoted to the history of the excavations. Not surprisingly, the history presented on the panels began in 1999, without a single reference to the previous 20 years of excavations in Orce.[41]

On the other hand, public mention of Gibert and the Orce Man at the press conference was unavoidable since journalists still remembered the controversy. In this context, Martínez-Navarro chose to state clearly that the cranial fragment known as the Orce Man did not belong to a hominid. He then compared the controversial nature of the previous Orce Man and the way it was presented (hastily and without precaution) with the undisputed nature of the new Orce Boy and the way it was presented. This new finding was presented more than ten years after it was discovered and, apparently, during that time the researchers had studied it in depth.

To sum up, the new discovery had to be precisely that: new, very new, with no ties to any other research or discovery related to Orce. Like Gibert rewriting his own history and recovering the past controversy in his popular science book in order to legitimate his position, Martínez-Navarro and Toro also wanted to write their own history. However, they wanted to write as if there were no past; voluntary amnesia to create a clean slate for the Orce research.

∴

Surprisingly, around twenty days after the publication of the article, the *Journal of Human Evolution*'s publisher, Elsevier, removed the online in press piece on the hominid tooth. Elsevier stated only "the publisher regrets that this article has been temporarily removed. A replacement will appear as soon as possible in which the reason for the removal of the article will be specified, or the article will be reinstated."[42] This temporary removal was not announced in Spain by the researchers or the politicians and, of course, it was not presented in a press conference. News of the removal came first from an online forum on prehistory called *Prehistoria y Vida* (Prehistory and Life). On Sunday, 17 March

41 El Niño de Orce 2013g

42 LaBaracA 2013.

2013, a forum user and administrator called LaBaracA announced the removal.[43] By 20 March, the news appeared in several blogs, tweets, online magazines, local newspapers, and also some English language blogs.[44] These places became the space for debate around the reasons behind the removal, the state of the dispute on palaeoanthropological research in Orce, and the relationship between the new finding and the old Orce Man controversy. In general, traditional national newspapers (which widely echoed the tooth's presentation) did not pay attention to the removal; it was mainly discussed online. The American blog *Retraction Watch*, which works "tracking retractions as a window into the scientific process", also commented on the removal.[45] According to *El Ideal de Granada*, Lluís Gibert kept "receiving phone calls from colleagues and the media" when the withdrawal was announced, showing the attention it received at the time.[46]

The role of new media in the dispute will be analysed in the next section, but it is crucial to state here that some of the primary information used next came from digital media. This kind of source was far more useful than traditional newspapers or institutional press releases. Therefore, it is necessary to bear in mind that what follows is mainly based on blogs, chat rooms, posts, and comments, and on personal communication with some of the actors involved.

After the publication of the article, at least four different scholars sent letters of complaint to the *Journal of Human Evolution*'s editor: Lluís Gibert; Enrique García-Olivares, the immunologist that performed immunology tests on the Orce remains in the 1980s; Francesc Ribot, a palaeontologist that had also worked with Gibert; and Mercè Piqueras, a microbiologist, science journalist, and blogger.[47] Apparently, the complaints were mainly about the absence of Gibert's name in the article: firstly, with regard to the discovery of Barranco León and its dating; secondly, in relation to the 1999 publication of a tooth fragment from the same site; and thirdly, with regard to the article's presentation of a stone tool as unpublished when it had in fact already been published by Gibert and his team in *Antiquity*, and, more surprisingly, the erasure of the

43 *Ibid.*

44 For instance: Ansede, Materia 2013b; El Niño de Orce 2013h; El Niño de Orce 2013c; Utrera, *El Ideal de Granada* 2013; Oransky 2013; Maju 2013; Piqueras 2013a; Piqueras 2013b; Facebook Cuarto Milenio 2013; and Rico, *Granada Hoy* 2013.

45 Oransky 2013.

46 "Luis Gibert no paraba de recibir llamadas de colegas y de medios de comunicación", Utrera, *El Ideal de Granada* 2013.

47 Personal communication with Lluís Gibert 2014.

tool's classification number.[48] To sum up, the main complaints were about the omission of Gibert and his team's research and the presentation of some of their work as if it were new. Apparently, these complaints addressed to the *Journal of Human Evolution* led to the temporary removal of the article. After the removal's announcement, Elsevier updated their statement by saying that "the reasons of the withdrawal resulted from a dispute with another team of authors who were unhappy their work wasn't cited."[49]

Online articles included statements from Bermúdez de Castro, Paul Palmqvist, Bienvenido Martínez-Navarro, and Lluís Gibert.[50] According to the online science magazine *Materia*, Lluís Gibert had achieved the removal of the article by sending a letter to the editor. Palmqvist stated that the editor told them to include some reference to Gibert and his research in the article. He also accused Gibert's son of acting out due to a "personal war" with the authors.[51] Martínez-Navarro said that the "editor has been pressured" to retract the article, and ironically stated that it was "not necessary to say by whom".[52] Bermúdez de Castro, one of the co-authors of the article and an expert in fossil teeth, handled Gibert's team's publication of the previous tooth from Barranco León by saying that the journal in which it was published, *Human Evolution*, was a "parish bulletin" that accepted anything with no proper scientific criteria.[53] On the other hand, Lluís Gibert stated that several scientific writings on the "archaeology, palaeontology, and dating of the site [Barranco León]" had been published before the new article and were not cited either.[54]

A lively debate took place in the comments section of the *Materia* online article. Some comments criticised the journalists for claiming that Gibert had the ability to have the article removed. Bermúdez de Castro's words were also commented on, ironically highlighting that both Palmqvist and Bermúdez de Castro had published pieces in the supposed 'parish bulletin' (*Human Evolution*) during their careers.[55] Furthermore, Palmqvist's article in that journal was precisely on the earlier tooth found by Gibert's team in Barranco León.[56] Some of the comments were made by scientists, journalists, or commentators

48 *Ibid.* See also El Niño de Orce 2013h or Piqueras 2013a and Piqueras 2013b.

49 Oransky 2013.

50 Ansede, *Materia* 2013b or Rico, *Granada Hoy* 2013.

51 "… guerra personal", Ansede, *Materia* 2013b.

52 "el editor se ha visto presionado para tomar esta decisión, no hace falta que diga por quién", Rico, *Granada Hoy* 2013.

53 "hoja parroquial", Ansede, *Materia* 2013b.

54 "… arqueología, paleontología humana y edades del yacimiento", Rico, *Granada Hoy* 2013.

55 Bermúdez de Castro/Nicolás 1996; Aguirre et al. 1990; Arribas/Palmqvist 2002.

56 Arribas/Palmqvist 2002.

involved in the story, such as Francesc Ribot, Mercè Piqueras, or El Niño de Orce. After these first accounts, the online dispute grew. The *Materia* article was reproduced in full and commented on in several blogs. On Twitter, El Niño de Orce started a specific dispute with Isidro Toro, an active user of this social network, thus showing that the blog was being followed and read by the story's main actors.[57]

Toro had changed over to new social media but other scientists with long-term involvement in the controversy, such as Enrique García-Olivares, still used traditional channels. García-Olivares sent an opinion piece to the Granada newspaper *El Ideal de Granada*.[58] He compared the Orce case and the omission of Gibert's name with George Orwell's novel *1984*, which was ironically the same year that the controversy started. According to García-Olivares, someone or something similar to Orwell's "Ministry of Truth" wanted to "vaporise" Gibert in order to impose a "single view" which was, according to García-Olivares, Atapuerca's single view.[59] Though it was published in a traditional newspaper, García-Olivares's letter also reached the online debate and was reproduced or commented on in some blogs.[60]

On 5 May, the *Journal of Human Evolution* republished the scientific article on the tooth with the title changed from 'The oldest human fossil in Europe dated ca. 1.4 Ma. at Orce, Spain' to 'The oldest human fossil in Europe, from Orce (Spain)', due to more detailed information on the dating that meant the tooth was only "estimated to be close to 1.4 Ma."[61] Beyond the new title, some small changes in dating, and the acknowledgment of Gibert in the image of the stone tools, the restored article's main novelty was a paragraph devoted to the remains claimed to be hominid by Gibert and his team. According to the new article, VM-0 and the other remains did not have "enough anatomical resolution to be ascribed to the genus *Homo*".[62] The article also commented on BL5-0, the other supposed hominid tooth fragment found in Barranco León. Toro et al. stated that Gibert et al.'s conclusion that the remain could be from a hominid "should not be accepted without caution" due to the "possibility that it could belong to a deciduous tooth of *Hippopotamus antiquus*, one of the large mammal species that are better represented in [...] Barranco León." The paragraph concluded, "the more parsimonious explanation (according to

57 El Niño de Orce 2013j.

58 García-Olivares 2013.

59 "Ministerio de la Verdad", "vaporizar", "pensamiento único", *Ibid.*

60 The letter was reproduced in full in Piqueras 2013c and partially in El Niño de Orce 2013e.

61 Toro-Moyano et al. 2013b, definitive version of the article. See also: El Niño de Orce 2013k.

62 Toro-Moyano et al. 2013b, 1.

Occam's razor) is that BL5-0 is probably part of a hippo tooth. Thus, tooth specimen BL02-J54-100 [the new tooth] [...] is the first human remain found at the Orce sites."[63] With the publication of the new article the dispute ended.

This is a short but significant incident that reveals how the debate around the famous Orce Man (VM-0) is perhaps over, but the dispute about by whom and how research is carried out in the town's sites is still open and lying dormant until the next 'episode'.

Although scholarly attention to the role of social media, or online media more generally, in scientific communication is constantly increasing, not as much expected works have so far dealt with it with completely. Despite the level of importance that this type of communication has reached, scientists' and science journalists' use of it still lacks attention from sociologists and science and technology studies scholars.[64] Yet, some studies have been devoted to the use in scientific popularization or within scientific controversies of specific online platforms such as Twitter, YouTube, or blogs.[65] In these few cases, most of the commentators acknowledge the potential of these new media to change scientific communication and to allow for a more open and 'real' dialogue between scientists and the general public.

Social and Digital media is increasingly being considered as part of both the scientific process of knowledge creation and the process of its dissemination, which this thesis has argued that cannot be separated. Scholars have highlighted for instance the role that platforms like Twitter could play as a post-publication review, as an extension of science communication or even as informal ways of legitimation of knowledge. Yet, at the same time, most of them also point out the difficulties. They highlight the often "dubious quality" of much of the information on the Internet and the low standard of peer-to-peer and professional-to-public discussions in forums like blogs, which easily "degenerate into name-calling or focus on trivial aspects of the issue."[66] In

63 Toro-Moyano et al. 2013b, 2.

64 Trench 2012, 273. See also Bucchi 2017. Some accounts include: Allan 2011; Bubela et al. 2010; or Cody 2010.

65 For Twitter: Veltri 2012, or Chew/Eysenbach 2010. For blogs: Trench 2012, Kouper 2010 or Colson 2011. For YouTube: Wellbourne/Grant 2015. For controversies see Yeo 2016 or Hopke/Simis 2015.

66 For post-publication review see Yeo 2016. For extension of science communication see Büchi 2017. For informal legitimation Brossard 2013. For critics see Bubela et al. 2010, 516–517 and Trench 2012, 286–287.

this sense, the Orce tooth debate 2.0, framed in the larger Orce controversy, reveals interesting aspects of how the advent of the 'digital revolution' has influenced the way that scientific discoveries and controversies are presented and discussed in public. An old pre-internet controversy that reappears in the online times can help us to identify similarities and differences between the different media contexts. The Orce Boy article's removal highlighted the potential of new media as a kind of non-professional review, while also helping us understand general trends that, despite changes in the characteristics of the media, are still intrinsically involved in the way that science works and scientific knowledge is produced and circulated.

As argued earlier, the way that the tooth was presented shows how, more than 30 years later, the need to have a public 'performance' in order to present new discoveries has not changed. Scientists announced the discovery to the media before its scientific publication in a ceremony with politicians and media representatives. The 'oldest European' label appeared again as a perfect way to 'sell' the Orce discovery. The intentions of the discoverers were very similar to those of Gibert, Agustí, and Moyà-Solà back in the 1980s, despite the fact that the social, political, and scientific situation in 2013 did not cultivate same level of media attention that the original Orce Man discovery received. The economic situation in Spain was not ideal for spending money on 'useless' science. Atapuerca's spectacular discovery in the 1990s and its broad media following also downplayed the significance of the new Orce findings. After the Atapuerca discovery of almost complete craniums and the naming of a new species, a single tooth, even from the earliest European, was an attraction for the Andalusian press but not really an issue for the national press, beyond a single half-page report.[67]

Yet, in the 21st-century case, while traditional media attraction was still necessary for getting public and political support, it was much less crucial in the way that the circulation of scientific knowledge and information among specialists and laypeople worked. Online forums, blogs, and comments on news pieces allow for a new form of knowledge communication that has apparently become faster, more far-reaching, and more dynamic. Immediately after the publication of the news piece in *Materia* on the 'temporary removal' of the *Journal of Human Evolution* paper, the piece's comments section displayed many responses to the content of the article. Some of the comments were linked to other pieces in blogs and newspapers that also echoed the news and thus amplified the distribution of these pieces. Likewise, reports on the discovery and

67 For Atapuerca see Hochadel 2013b.

the removal were disseminated on Twitter. Some of these accounts, like the *Materia* piece, were reproduced several times on blogs and other sites. Some scientists and journalists involved in the affair participated in the diffusion and made comments. One example is Francesc Ribot, a palaeontologist who was a former collaborator of Gibert's. Ribot made comments on the *Materia* piece, the *El Niño de Orce* blog, and Facebook pages, and criticised the authors of the article for not including a reference to Gibert and the previous tooth from Barranco León, including technical details about Gibert's findings in his critiques. This highlights how online media, such as blogging, can be especially influential in controversial and public issues, providing fertile ground for lively discussions.[68] In addition, this online discussion could reach the 'real' world and influence the way that researchers present discoveries, as happened in the presentation of the new tooth to the people of Orce.

Like comments, Twitter and (especially) blogs allowed the parties involved to express themselves directly to their 'publics'. Isidro Toro was very active on Twitter where he shared news about the discovery. José María Bermúdez de Castro also commented on the discovery on his blog, and Mercè Piqueras, who sent one of the letters to the *Journal of Human Evolution*'s editor, echoed the removal twice on her blog.[69] Yet, most of the scientists involved did not directly use their own blogs but instead used comments sections to discuss the removal, and a lively debate about Orce took place particularly on two or three forums. Therefore, in the 2013 Orce case, online digital media became the space where scientific knowledge was presented by scientists themselves, as well as being a forum for discussion between them and with the general public. For at least five days between the press conference and the publication of the online article, the only source of information on the new Orce tooth discovery was the media. Both traditional and online media echoed the scientists' statements from the presentation and featured new ones. Blogs, Twitter, and Facebook accounts spread this information, which even reached some English-speaking and international newspapers and online media.[70]

At the same time, these new digital media, like the traditional media before, were the main sources of information for those interested in the issue. Since the presentation of the discovery to the press took place before the scientific paper was available, scientists' statements in the press conference and

---

68 Trench 2012, 276.

69 Bermúdez de Castro 2013; Piqueras 2013a, and Piqueras 2013b. See analysis and examples of scientists' use of blogs in Colson 2011.

70 Newspapers: Fotheringham, *The Independent* 2013, and Marot, *der Standard* 2013. Blogs entries: Theolduvaigorge 2013; Maju 2013; Jones 2013. See also: El Niño de Orce 2013l.

subsequent comments, blogs entries, and online newspapers and magazines were the only source of scientific information on the tooth, at least until five days later when the article was accessible online in press. Then, the article was only accessible paying 19.95 dollars or to those affiliated to an institution with access to the *Journal of Human Evolution*. Therefore, the public again had difficulties in accessing published scientific information on the tooth while being very easy to get information in Twitter or blogs. During this period without access to the article, an English-language blog called *The Olduvai Gorge* reviewed it, using some paragraphs from the original and even published an entire image from it.[71] This image was also reproduced on the blog administrator's Twitter account and it appeared on the *El Niño de Orce* blog, thus reaching interested members of the Spanish general public that did not have access to the actual scientific article.[72] This shows how information from scientific articles can easily flow to blogs and other online media that become direct sources of information for the interested general public.

Somehow, this situation was similar to the three years that passed from the 1984 *El País* front cover to Agustí and Moyà-Solà's article in 1987, when information about the famous Orce Man controversy was mainly disseminated in newspapers. This gives the media a crucial role in the way that scientific knowledge circulates both among the public and among the scientists involved, precisely during times when the issue at hand receives the most attention. It seems then that as well as the fact that science's need to have a public presence is similar to the 1980s case, with new media it is much easier and faster for the parties involved to 'go public' and 'publish' an opinion in a blog or a comment on a news piece. This confirms the often-claimed democratic nature of new online technologies and platforms. In addition, this situation also allows for better analysis of the public prevalence of opinions and the dominance of one opinion over another. These new technologies allowed voices as Ribot's to be heard in the debate, voices that might have been silenced or barely noticed in more traditional media. The Internet and social media allow the analyst to 'listen' to the voices of these different 'publics'. What was very difficult in the 1980s becomes almost one of the most interesting sources in the 21st century.

Later, an online prehistory forum was the first to announce the temporary removal of the article; a scoop that was distributed online that same day and later reached magazines and newspapers that joined the discussion. Here it is crucial to highlight the relationship between the online public discussion

71 Theolduvaigorge 2013.

72 El Niño de Orce 2013a.

about the article's removal and the private discussion-taking place between the editor of the *Journal of Human Evolution* and the scientists involved on both sides. This discussion was mainly about what role Josep Gibert and his team's research in the Orce area played and how they should be acknowledged or presented in the publications of the new team. The direct involvement in this private discussion of the scientist, scientific journalist, and blog writer Mercè Piqueras, together with the detailed entries published on the blog *El Niño de Orce* and the article authors' comments to the press, allowed this private discussion to be opened to the public.[73] In a way, two separate discussions cannot be considered, one public and one private, but one single discussion with a public and a private side.

Finally, as Jim Secord argued in his book *Victorian Sensation* for the 19th century, the introduction of new press technologies, combined with an increase in literacy and the development of railway communications, allowed publishers and authors, like Robert Chambers author of the *Vestiges*, analysed in Secord's work, to produce a new science product: the first popular science books, which changed scientific communication.[74] Moreover, these new technologies also changed periodicals and newspapers increasing circulation of knowledge and audiences. This, of course, affected the generation of scientists that grew up with these changes. For the likes of Huxley or Tyndall, less specialised publications became central in their professional agendas.[75] Similarly, early 21st-century communication media, increasingly used by many different publics, are changing the way that scientific knowledge circulates and are therefore modifying scientists' actions. The brief dispute between the authors of the tooth article and the defenders of Gibert's legacy shows how new media facilitated faster communication and the more direct involvement of scientists, who could express their opinion on the dispute. This also led to a wider distribution of these comments and of the information on the case more generally, allowing for all interested audiences to be addressed and not only the readers of certain newspapers and journals. The temporary removal controversy also shows how scientific discussion among peers and with the public took place online. This discussion used direct, detailed, and high-quality information on the issue from both sides of the controversy. Long articles in *Materia* and posts on *El Niño de Orce* are clear examples of this informed discussion. This

73 See: El Niño de Orce 2013h; El Niño de Orce 2013m; Piqueras 2013a and Piqueras 2013b. An especially intense debate in: El Niño de Orce 2013c; Ansede, *Materia* 2013b; or Facebook Cuarto Milenio 2013.

74 Secord 2002, 2 and 167.

75 Lightman 2011.

allowed the interested public to access valuable information from both sides of the controversy, which in turn allowed for informed positioning on one side or on the other. Blogs, but also Twitter and comments sections, could be the new 'science products' that nowadays scientists will use more and more often in their scientific debates and therefore in their scientific practice. Beyond the distribution of knowledge, these new products could also be used to legitimate a certain position. Previous online presence (for instance, managing a blog with regular and influential readers or being active on Twitter and having sufficient followers) means having power in a possible forthcoming controversy since your accounts or comments will have wider distribution and visibility. Like popular science books, blogs and Twitter could be increasingly be both, part of scientist's agendas, and, when a dispute starts, 'enlarged battlefields' where scientific issues could be discussed and controversial ideas put forward without the usual controls of scientific publications.[76]

To sum up, both in the 1980s and in the early 21st century, scientific positions and critiques regarding scientific issues had to be public first in order to be relevant. The strategies and aims of the public dissemination of a new discovery apparently have not changed much. Scientists made similar claims in similar ways. Yet, the circulation of the information is now different. The presentation of the second discovery was not as widely followed as the first and the temporary removal dispute was brief, but this analysis could reveal new trends in scientific communication. Blog writers and even journalists that wrote relevant pieces on the issue, which were commented on and shared, were much more informed and more implicated in the story itself. The immediate consequence of this is that what years ago was a private opinion that could potentially have its own public side via newspapers, can now be more immediately public through, for instance, blog posts. Somehow, according to this dispute analysis, online communication reduces the distance between the space of the generation of scientific knowledge, the space of discussion, and the space of circulation and diffusion. Apparently, the gap between the so-called scientific community and what may be call an interested public becomes smaller. Back in the 1980s, newspapers could act as channels for political and public support and fund raising, as a stage for scientific debate among specialists, and as a place for scientific validation together with scientific forums.

Now, online communication (ranging from online newspapers, Twitter, and Facebook, to blogs and comment sections) is, increasingly, inside scientific debate itself. As the Rising Star Expedition case showed at the beginning of this

76 See of these ideas applied to science books in Hochadel 2013c.

chapter, the publication of scientific articles will then be just another step, and not necessarily the first, in scientific communication. The interested general public could then be involved in this communication more than ever before. Presumably, scientists, like Western society in the early 21st century as a whole, will develop, discuss, circulate, and validate their practice and knowledge more and more online, and, therefore, in an increasingly public manner. This could bring with it several problems about the validity and final acceptance of scientific knowledge and the possible ‘danger’ of less ‘rigorous’ knowledge. Yet, at the same time, if managed well, the online presence of scientific communication could lead us towards more open and democratic scientific practices negotiated among and between a wider range of easily and more informed ‘publics’. Therefore, debates, discussions and decision making on scientific issues, and on funding issues, could take place not only in the closed traditional spaces, in the scientific ivory towers of conferences and specialized journals, but also, after or in a parallel way, in the online public spaces. Online public spaces where these informed ‘publics’ could participate in the communication and development of a new democratic scientific culture.

CHAPTER 7

The Orce Man: Controversy, Failure, Media and Politics

With this book, I wanted to provide an account of the Orce Man controversy, specially of the first years, as complete and detailed as possible. From its discovery to the *El País* front cover, from the 1995 conference to the 2013 'Orce Boy', I have tried to put together the whole story without taking sides and without taking for granted any supposed truth. Through the Orce Man example, I also wanted to contribute to the understanding of particular periods in the recent history of Spain and the role played by science in them. The Orce Man story is also useful to fulfil a piece of the history of palaeoanthropology, between the 1970s great African discoveries and the early twenty first century controversial findings. At the same time, I tried to improve the understanding of scientific controversies and especially public scientific controversies in this recent period of history of science. The strategies and 'tools' used by the actors involved and the paths that knowledge takes in these cases have been under analysis throughout. With this detailed analysis, the Orce Man story has emerged as much more complex than in previous accounts, thus showing how it could be useful for better understanding palaeoanthropological scientific practice and its relationship with the media and with politics. Finally, the brief description of Niède Guidon's and Pedra Furada case has shown that, longer or briefer, bigger or smaller, this kind of controversial relationships between scientists, journalists and the public is common in late 20th century scientific practices.

In the introduction to their book *Scientific Controversies* , Machammer, Pera, and Baltas stated that scientific controversies that are focused on the result or methodology of one single experiment or on the interpretation of any scientific result or object often hide disagreements on more general theoretical conceptions.[1] In the Orce case, there is one single point of contention: the classification of the cranial fragment VM-0 as a human species, *Homo* sp., or its classification as a member of the genus *Equus*. This was also related to the fossil's dating, which could potentially allow for the placement of the crucial label 'First European'. Yet, this labelling of the Orce remain did not cause any conceptual disagreement between the involved scientists at the beginning of

1 Machammer/Pera/Baltas 2000, 8–9.

 | DOI:10.1163/9789004431508_009

the story when there was no controversy. Moreover, the main general theoretical claim made by Gibert (that Orce showed very early hominid presence in Europe) was shared by most of his later opponents (the de Lumleys, Agustí, the Atapuerca trio, Martínez-Navarro…), who acknowledged that stone tools found in the Orce area proved this claim. Even foreign scientists, like Roebroeks, who first supported the so-called 'short chronology' of the occupation of Europe (in which Orce did not fit), later admitted that this did not apply to southern European cases, including Orce.[2]

Nevertheless, there was another claim that was debated. It was the claim that these early hominids crossed, over one and a half million years ago, the Strait of Gibraltar to reach Europe.[3] As previously noted, the most accepted route for the hominid 'journey' to Europe is the Levantine Corridor, which was further strengthened with the 1990s findings in Georgia. The Atapuerca researchers, for instance, maintained (in both scientific and public forums) that they were completely against the Gibraltar claim, and despite the geographic closeness of Africa and Spain, they argued that Atapuerca hominids came all the way from the East to the Iberian Peninsula. Yet, there were also some international researchers, such as Soondar or Roebroeks, who supported the idea of an early hominid crossing via Gibraltar, or at least found it plausible. Roebroeks, for instance, by admitting the possibility of an early hominid presence in the Iberian Peninsula but not beyond the Pyrenees, was also admitting the arrival of these hominids through Gibraltar. Agustí acknowledged that the arrival through Gibraltar was a clear possibility and even discussed the issue in one of his popular science books. Meanwhile, for Moyà-Solà this was a pointless debate because it would never be totally proven and therefore it would never be a scientifically stated fact.[4] In short, the Strait of Gibraltar crossing was not a clear debate between the same parties that disputed the Orce Man's validity. Therefore, it appears that the controversy over the cranial fragment did not hide any crucial disagreement over a more general theoretical issue.

Instead, as this book has tried to show, the controversy over the cranial fragment hid issues like personal or political disagreements, economical interests and efforts to gain public prominence. The way that Gibert was excluded from the Orce area excavations was based more on political and personal factors than scientific ones. Gibert's way of acting, his emphasis on the cranial

2 Dennell/Roebroeks 1996.

3 Gibert/Gibert/Iglesias 2003.

4 See Arsuaga's statements in TVE 2011; Carbonell's in Guàrdia/Pou 1996 and see also Carbonell/Estévez 1983, 152(68); Carbonell/Sala 2000, 113. Dennell/Roebroeks 1996, Agustí/Lordkipanidze 2005 and for Moyà-Solà statements see Garcia Ruiz 2003.

fragment, his highly public profile, and even his political ideas played a crucial role in the controversy, in his isolation, and in the way other scientists criticised him and classified him as a 'bad scientist'. These factors often played a bigger role than specific scientific claims. In many ways, the controversy was not about whether VM-0 was a hominid or not, but about by whom and how the evidence related to the bone was presented to both scientists and the general public. Somehow, deciding whether the cranial fragment was a hominid or not was the same as deciding whether Gibert's methodology was correct or not, whether he had enough credibility as a scientist or not, and whether he was a valid expert or not. As Steven Shapin stated, "credibility and the validity of a proposition ought to be one and the same".[5] Seen in this way, the Orce Man controversy could be, following Aristides Baltas's classification in the same book *Scientific Controversies*, a "surface controversy", in which what was at stake was not a scientific problem but a way of doing science. The 'boundary work' analysis discussed earlier also supports this approach to the controversy.[6] As Richmond stated about the Taung Boy case, the discovery of this possible human ancestor was not only that, "but also a new opportunity for Dart to fulfil his scientific ambitions. The debates over the credibility of his claim were, in this way, also debates over Dart's professional standing in science."[7]

Likewise, at some points this controversy was also a fight for a scientific or even a public 'niche'. After Miquel Crusafont's death and with the Orce Man generating scientific media hype, Gibert became not only the new director of the *Institut de Paleontologia de Sabadell* but also a Spanish public expert on palaeontology and palaeoanthropology. Gibert became a 'visible scientist' in Rae Goodell's terms, a scientist that had an interesting and appealing topic to 'sell' to the public and to politicians and therefore to get funding.[8] At that point, Gibert did not present any controversial issues or generate controversy among the scientific community despite, for instance, presenting in public research that had not been published in scientific literature. Somehow, Gibert established himself as the 'Spanish public palaeontologist', the expert that the media could call on when necessary. This situation may have aroused distrust among peers.

Once the controversy broke out, those same peers felt that Gibert, and his insistence on the validity of the Orce Man, were an easy target to direct their attacks in order to eliminate him as a rival for the 'Spanish public palaeontologist'

5 Shapin 1995, 255–56.
6 Baltas 2000, 46.
7 Richmond 2009, 3 and 23–24.
8 Goodell, 1977

position, and therefore in the fight for funding and popular praise. Several such attacks were directed at Gibert and he was substituted as director of the *Institut*, mainly because of the controversy's strong media presence.

After that, Gibert became a different kind of 'visible scientist': an outcast marginalised by his scientific peers. Following Goodell's analysis, this situation led him to brake some theoretically "unwritten rules" of science, like avoiding public controversy.[9] As Goodell states, "for unpopular scientists whose views are criticized or ignored by fellow scientists, the media may represent a chance for a more sympathetic forum."[10] This situation, added to the political dimension explained in part three, prevented Gibert from continuing to work in Venta Micena and in the Orce area more generally. Control over the excavations was also a 'niche' to be fought for. Gibert, despite being very popular in Orce, was not welcomed by the *Junta* politicians or scientists, or scientist-politicians, who prohibited his work as much as possible.

In their book, Machammer, Pera, and Baltas also point out how disputes between scientists can foster the development and growth of science, bringing innovation into scientific practice and generating opportunities for junior scholars.[11] In the Orce case, on the one hand, it was the early uncontroversial discovery that propelled the *Institut* into a new age of scientific research and allowed the scientific collaboration between Andalusian and Catalan institutions. The subsequent controversy led to Gibert's isolation in the *Institut* and the impossibility of continuing scientific research in the Venta Micena site, which was closed and had no scientific activity for many years. On the other hand, the desire to demonstrate the hominid nature of the cranial fragment also brought new and innovative techniques to palaeontology and palaeoanthropology in Spain. The immunological test that attempted to identify human proteins in fossils was performed for the first time in Europe on the cranial fragment and on other remains from Orce. The mathematical fractal analysis performed twice by Paul Palmqvist on drawings of the Orce bone was also a very innovative technique applied to the analysis of palaeontological remains.

Furthermore, the Orce Man controversy created great opportunities for young scholars like Bienvenido Martínez-Navarro or Gibert's son, Lluís, and other scientists linked to Gibert's group who based their early careers on the Orce controversy. They published articles and books, organised conferences, and based their doctoral theses on work related, in one way or another, to Orce. In a way, they grew up as scholars within the context of a scientific controversy.

9 Goodell, 1977, 92/130.
10 Goodell, 1977, 180.
11 Machammer/Pera/Baltas 2000, 10/30.

Later, when Martínez-Navarro changed sides, he was able to apply for jobs and present excavation projects thanks to his extensive curriculum vitae based on several national and international publications on the Orce Man and Orce fauna (his speciality) and on his participation as part of Gibert's team in conferences, research stays, and excavation and research projects. In 2003, he got a permanent position from the Catalan government as an ICREA research professor at the *Universitat Rovira i Virgili* in Tarragona and later at the *Institut Català de Paleoecologia Humana i Evolució Social* (IPHES), founded in 2004 by Eudald Carbonell, who was also the director.[12]

To sum up, the scientific and public controversy of the Orce Man paralysed the excavations in Venta Micena and left Gibert with very little financial support from the institutions. Yet, it also resulted in the application of innovative trans-disciplinary techniques in palaeontology as well as creating academic and professional 'niches' for young scholars that were involved in the controversy very early on in their careers.

It is not a risky assertion to state that hundreds of articles have been written on the controversial Orce Man bone fragment. Most of them presented evidence that supported either its classification as a hominid or its classification as an equine, or some other animal.[13] The controversy itself little by little vanished and an almost complete agreement was reached among experts, despite a minority of scholars still claiming that the bone belonged to a hominid. It could be argued that a consensus among the vast majority of the scientific community was reached against the Orce Man bone fragment's hominid classification. Yet, the concept of consensus implies agreement between the two disputed parties, and this never happened in the Orce Man controversy. Gibert and his team were gradually overpowered until his death but never accepted the equine classification of the bone. The supposed consensus reached was more a matter of domination and exclusion than that of agreement. Considering this situation, instead of explaining the eternal question about the Orce Man (if it is a man or not), I will try to understand why, in fact, this question is impossible to answer. In order to tackle this, the concept or idea developed by Harry Collins of the 'experimenter's regress' or in this case the 'palaeoanthropologist's regress' will be used.[14]

In his study of experiments for the detection of gravitational waves, Collins revealed a kind of vicious circle in this detection. In the scientific community studied by Collins, there was no consensus over the actual existence

12 See Martínez-Navarro's curriculum in Martínez-Navarro, n.d.

13 More details on the features used for each attribution see Annex 1.

14 See Collins 1985 and Maerker 2006.

of gravitational waves. In addition, the experimenters that first worked on this issue did not know what correct detection of these waves should look like. This led to a situation in which the result of an experiment was interpreted according to whether gravitational waves existed or not for a certain experimenter. Therefore, depending on the experimenter, a competent experiment was either one that detected the waves or one that did not. Therefore, the experiment itself did not have any element that allowed for a decision on whether it was in fact a good way for detecting radiation or not. This is what Collins called the 'experimenter's regress'. He concluded that the criteria for accepting a certain experiment must be found outside of the experiment itself.[15] Similarly, Anna Maeker applied the 'experimenter's regress' to a natural history case, establishing the 'taxonomist's regress', in which taxonomic classification, like in the Orce Man case, is what was at stake.[16]

When the Orce Man bone fragment was found in the 1980s, there were no 1.5-million-year-old hominid craniums to compare the Spanish remains with. When the controversy emerged, researchers from both sides had the opportunity to compare it with ancient *Equus* remains, modern horse and ass bones, and even modern human remains, but none of them had the chance to compare a supposed 1.5-million-year-old hominid with a real 1.5-million-year-old hominid. Only later, Martínez-Navarro compared it with Kenyan and Javanese remains with similar dating, although not exactly the same.[17] Therefore, when the controversy broke out, the classification of the Orce bone fragment as a hominid or an equine hinged on the presence of the inner crest in 1.5-million-year-old hominids, and there was no other fossil that could help the researchers decide whether those hominids had that concrete feature. So, there was no way to establish or refute this claim (that 1.5-million-year-old hominids have, or not have, a crest on the occipital bone) other than through consensus among the scientific community. Thus, as this consensus did not exist, the classification would always be dependent on the position of each researcher in the controversy. Therefore, no act of classification was ever fully correct for all parties. The classification of the specimen will not depend on the specimen itself but on elements external to it. This is Harry Collins's 'experimenter's regress' applied to the Orce Man controversy.

Some may say that, outside of the controversy, there must be someone or something that could act as a judge to decide finally who is right and whether these hominids had a crest or not. Yet, when in a controversy, what is at stake

15 Collins 1985, a more recent account of this situation in Collins 2017.

16 Maerker 2006.

17 Martínez-Navarro 1996.

is not only whether the bone is from a hominid or not, but how is this decided, the criteria by which this is decided, everybody involved in the debate has to take part in deciding what the rules are for defining what a 1.5-million-year-old hominid looks like. As Pierre Bourdieu argued, "inasmuch as the definition of the criteria of judgment [...] is itself an issue in a struggle, there are no good judges, because there is no judge who is not also party to the dispute."[18]

Therefore, when there is no previous research or remains that could act as a reference for all the parties, there can be no external judge, as this judge's decision could also place him on a particular side of the controversy. Therefore, with no reference and no judge, the scientific consensus between rival parties is impossible, since the controversy cannot be solved using merely rational factors. Since Gibert and his team insisted that the Orce bone was in fact a man, and de Lumley first and Agustí and Moyà-Solà later insisted that it was a young ass, consensus could not be reached and it was an argument that could not be solved through traditional scientific procedures. It could only be solved through social factors or not at all. As Collins also argues, the experimenter's regress is solved through 'social and cultural networks' of the involved actors and objects, such as the credibility of the experimenter or the institution in which the experiment is performed.[19] These 'social and cultural networks' of the Orce Man that have been analysed throughout this book.

Then, how did the Orce Man controversy finally end? Did it end at all? In his contribution to the 1987 *Scientific Controversies* book, Tom Beauchamp characterised up to five ways that controversies can end. The first three ways ('sound argument closure', 'consensus closure', and 'procedural closure') deal with the rational imposition of consensus between disputed parties, which was not possible in the Orce case. The fifth one ('negotiation closure') does not require a fully satisfactory consensus for both parties but involves a negotiation between them with a compromise as its outcome, which did not happen in the Orce dispute. 'Natural death closure', Beauchamp's fourth proposal, is perhaps the most appropriate for the Orce case. In this mode of controversy closure, according to Beauchamp, the following are possible: "interest in the issues disappears"; "no new arguments are found"; or perhaps "combatants have tried" and the controversy ends due to "a dynamic flow of events" and "either is never resolved or is recast in the form of some related [...] controversy." Also in *Scientific Controversies*, McMullin offers a classification of how scientific controversies end in which the third kind of closure ('abandonment') may coincide

18 Clemens 1986, 426. See also Gross 1996, 8.

19 Collins 1985.

with Beauchamp's 'natural death closure'.[20] In the Orce Man controversy, until the year of his death (2007), Gibert and his team were still presenting new evidence (both in scientific and in public forums) that the famous cranial fragment found 25 years earlier was a hominid remain. After his death, the situation changed and these types of claims were no longer made. His son Lluís moved onto other issues and kept on fighting to obtain excavation permits in the Orce area, but he stopped trying to demonstrate the validity of the original cranial fragment. Some popular articles were published on the controversy after Gibert's death, but in general media attention for the case diminished substantially.[21] The processes of exclusion of Gibert from Orce's research finally worked. It was then the 'dynamic flow of events' that ended the controversy which was both 'never resolved' and somehow also 'recast' in Gibert's son's disputes and in subsequent events. With no possible rational consensus between the parties involved, the specific controversy around the original VM-0 cranial fragment was never resolved. It simply ended with the 'natural death' of one of its most active parties: Josep Gibert i Clols.

∴

During the second half of the 1990s and throughout the 2000s, the *Institut de Paleontologia de Sabadell Miquel Crusafont*, now renamed as the *Institut Català de Paleontologia Miquel Crusafont*, appeared again in all the media due to a series of discoveries of ancient great apes that were announced as the possible common ancestors of modern great apes and humans in Catalonia.[22] The leader of the discovery team was Salvador Moyà-Solà who was also among the authors of the scientific papers on the discoveries, some of which appeared in top journals such as *Nature* or *Science*.[23] These discoveries did not receive as much media attention as the Orce Man or the Atapuerca remains, perhaps because great apes are not as appealing as hominids.[24] Yet, they demonstrate the position of the *Institut* as a leading palaeontological institution in Catalonia, in Spain, and internationally.

These quite public discoveries have certainly helped to reinforce the *Institut*'s position in the last twenty years. Likewise, in the early 1980s, the Orce

---

20 Beauchamp 1987, 33 and McMullin 1987, 81–82.

21 Cardeñosa 2007 and Campillo 2010.

22 See, for instance: Redacción, *El Mundo* 1996; Antón *El País* 2005. For detailed accounts of the story: Moreno 2010, see also Hochadel 2013d, 292–293.

23 For instance: Moyà-Solà/Köhler 1996; Moyà-Solà et al. 2004.

24 Hochadel 2013b, 199–201.

Man discovery helped to re-launch the *Institut*, which was declining due to Miquel Crusafont's ill health and the end of the dictatorship. The coincidence of the consolidation of the new democracy and the Orce Man discovery 'saved' the *Institut*, which received new employees and funding. Until a year after the discovery's presentation (when the three researchers went to Paris and *El País* published the front cover that started the controversy), the Orce Man had been very successful in attracting public attention and funding. Both early popularisation and grand claims are common when new findings are presented to society. When a discovery is made, popularisation and supposed exaggeration are crucial in order to construct a whole machinery and narrative around the discovery to ensure popular support and funding. From this point of view, the beginning of a supposed failure story such as the Orce Man is no different in its core from a success story like Atapuerca. Usually, in most of the historical accounts in palaeoanthropology written by scientists themselves, success stories are a result of great endeavours, luck, scientific pursuit ... while failures, controversies, errors and hoaxes, are often explained by blaming the intervention of politics and ideology or non-rational and interested scientists.[25] This kind of accusation have been often made to Orce and Gibert. This is why comparing Orce with Atapuerca is so useful. It reveals how both success and failure could have similar origins, which leads us to think that these contextual issues, politics or interests, must be taken into account in a similar manner in both cases at not explain success only as magnificent 'adventure' stories and leave failure to irrational and obsessed scientists.

The *El País* front cover changed everything. Gibert tried to defend his position by adopting the discourses just mentioned, while Moyà-Solà and Agustí did not appear much until some years later. This allowed them to separate their name from the Orce Man, which became mainly a Gibert issue. From then on a controversy started over the taxonomic classification of the famous cranial fragment. Due to the characteristics of the dispute (mainly the inability to decide whether the Orce fragment was a hominid or not and the impossibility of a consensus since both sides maintained their claims) the classification became impossible to solve by scientific channels and became solvable only by social ones. This shows how the Orce Man failure story, as also happens with success stories, must be considered within its social and political context and in relation to the media.

The book has tried to analyse the main discourses used by both sides of the controversy in their dispute, not only for and against the Orce Man, but also

25 Van Reybrouck 2002, 160.

in their struggle for resources, credibility, and prestige. On the one hand, the defenders of the Orce Man developed a discourse based on suffering and marginalisation that sought legitimisation through the use of the history and philosophy of science. In short, again, the Orce Man was a revolutionary discovery that changed the Kuhnian paradigm and therefore caused the rejection of the scientific community, which resulted in the lack of resources. The Orce Man had a series of 'enemies' within the scientific community that again reveals the geographical dimension of the case: at the Andalusian level, Pascual Rivas and Isidro Toro blocked permits and funding; at the Spanish level, the Atapuerca team dominated Spanish research, as well as funding; and, internationally, Henry de Lumley represented the establishment power that rejected Gibert's claims. This discourse was repeated in all kinds of forums from scientific papers to the popular press and popularisation books.

Yet, at the same time, any positive comment from a member of the very same scientific community was included and given a prominent place in the discourse. Phillip Tobias, Yves Coppens, and Emiliano Aguirre's support of Orce were used several times as another 'tool' in the controversy. This concept of 'tool' has been of significant help in this analysis, especially in chapter three. In this regard, it is necessary to highlight the importance of the materiality of the remains found and the sense of 'possession' and 'control' that often accompanies these discoveries. The display of the remains or the ceremony of bringing the fossils back to Orce show how these remains were used as part of the same rhetorical discourse.[26]

Finally, the Orce Man could be a failure completely rejected by the scientific community and therefore have little scientific relevance, but at the same time have social relevance through the early popularisation and controversy that incites the scientists' desire to control it. The Orce Man was then a scientific object that had several interpretations from the different actors involved. Even today, while for the majority of Spanish and international palaeoanthropologist do not consider VM-0 in their compilations of human ancestors remains, in the small village of Orce, the famous bone is still the major attraction of the local museum, were it is presented as 'the Orce Man'.

On the other hand, those scientists opposed to the Orce Man used a strategy of 'exclusion' type boundary-work in their treatment of Gibert's research. As already mentioned, this strategy consisted of categorising Gibert as non-scientific, non-rigorous, and non-serious in order to situate him and his team not only outside of the Orce research but also outside of scientific practice

26 More on the 'travelling bone' during the early years of the story see Annex 3.

itself. This strategy appeared several times during the controversy when things like funding, prestige, power or control of the bone and the site were at stake. Scientific dispute took place in the media. This situation led three palaeontology professors to re-establish the separation between science and the media by discrediting Gibert and his procedures. These professors, and later also the Atapuerca researchers, not only wanted to decide whether the bone was a hominid or not, but also whether Gibert's research was credible or not, and even whether Gibert was valid as a scientist or not. The boundaries that these scientists wanted to build placed the Orce Man outside of science, but also the controversy and the researchers involved, especially Gibert. Thus, this book has shown how power among certain sectors of the so-called scientific community works. A power that had the capacity to decide whether the Orce bone was a hominid or not and whether Gibert's methodology was 'science' or not. Gibert's persistence and the media attention received by the controversy obliged these sectors to show their authority in public and therefore the Orce Man's apparent success turned into failure, with the loss of funding and permits.

Since the very beginning of the discovery of the Orce Man remain, it has been in close relation to politics. Its early presentation in Granada, the later presentation in Sabadell, the *Diputació* funding of the palaeontological research, the political involvement in the Orce Conference or the 1998 debate around excavation permits in the Andalusian Parliament show how this relation has been very profound and intense during the whole controversy. The public presentation of the Orce discovery was a prominent event in the early democratic Spain, in which science had to play a very important role. This political context of state building is significant not only because the historian now knows the processes that were taking place in Spain at the time, but also because the actors involved in the story also knew about these processes. Scientists, but also politicians and journalists, knew that Spain was going through a profound change that could involve benefits for them. Recently, several historians of science of Franco's dictatorship are arguing that during this period scientists were not only working under the dictatorship, but also were a very active and important part of it, and of its legitimation.[27] Here I argue that science during the new state building process after the dictatorship not only operated in this context, but also became a very relevant part of it, helped to build it and its discourse.

27 See for instance Camprubí 2014 or Florensa 2017.

In addition to this political presence, at the beginning of the story the particular political context in Andalusia allowed the Orce Man to be used in the championing of the Andalusian identity, linked to the poor economic conditions of rural Andalusia. This political context of the early 1980s, with Spain entering a new democracy, clearly played a role in the way that the Orce Man was presented to society. This allows us to go beyond palaeoanthropology, even to go beyond the history of science, and to contribute to picturing a specific historical period and the way that public science worked in it.

Later in the controversy, excavation permits and funding for the Orce research became a recurring issue among Andalusian politicians who used it in their entirely political disputes (figure 22). In Andalusia, Orce became a completely politicised issue that caused confrontation between opposed parties on several occasions. To show the extent to which the Orce research was related to Andalusian politics, it is easy to recapitulate the five heads of the *Junta*'s Department of Culture that have been actively involved in the Orce issue and its presence in the media: Rafael Román participated in the press conference for the discovery's announcement; Javier Torres Vela responded to the media when the controversy broke out; José María Martín Delgado attended the closing session of the 1995 Orce conference; Carmen Calvo received the Orce bones from Gibert in a ceremony in Granada; and Luciano Alonso presented the Orce Boy remains to the press.

Finally, the highly politicised environment that surrounded the Orce Man issue in its different phases is useful for understanding the use that both scientists and politicians make of the public interest in science. On the one hand, scientists took advantage of this situation. Before the media, and side by side with politicians, they claimed that more funds were needed to improve their research and institutions. Their popularisation of the discovery was closely linked to this 'lack of funds' discourse which often achieved its aims. On the other hand, politicians from completely different ideological positions appropriated the Orce Man in order to reinforce their institutions and their own management of those institutions. The clearest example of this is the *Diputació*'s position when the discovery was presented and its desire to strengthen the *Institut de Paleontologia de Sabadell* as a scientific institution. All these political ambitions to 'appear in the Orce photo' contrast with the recurrent discourse to leave science to the scientists and only offer institutional support. In the Orce Man story, political interests not only affected science but also completely guided and shaped it.

FIGURE 22 A cartoon in *La Opinión de Granada* on the Orce research. The cartoon speech bubble reads 'At this site they're no longer looking for fossils, only political corpses'.
FROM BONET, *LA OPINIÓN DE GRANADA* 2007

Probably not long after January 1994, Josep Gibert received an undated letter written with a blue marker upon an article from *El País* entitled 'Lumley dismisses the Orce Man' at his office in the *Institut de Paleontologia de Sabadell* (figure 23). In it, a painter, which name has been erased, told Gibert that he was "uneducated in the discipline", but that considering what some distinguished Frenchmen said about Altamira, the opinion of this other Frenchman did "not deserve" his "sympathies".[28] This letter is one of the few sources that survive capturing how a member of the public received and processed information about the Orce Man in order to shape their own opinion on it. The painter most likely had read or heard about the French rejection of Altamira, maybe from Gibert himself or maybe from a different kind of source. When he encountered this news piece, he decided to support Gibert using the same discourse that Gibert also used, showing that these kinds of narratives really worked once they reached the public.

28 'Lumley descalifica al Hombre de Orce', Antón, *El País* 1994; "inculto en la disciplina", "no merecen mis simpatias" AJG-ICPMC: Letter to Josep Gibert, n.d.

Moreover, this single example reveals several of the points regarding the 'publics' of science. The way that the Orce Man was presented and popularised in the Spanish media and the way that the later controversy took place crucially shaped the character of the Orce Man as a scientific object. In this case, scientists presented brand new scientific claims to the media, journalists made big headlines out of the dispute, and cartoonists used it in their satires. In several phases of the Orce Man controversy, these kinds of communicative actions were as relevant as scientific communicative actions such as conference presentations, museum displays, or scientific papers.

Like conference presentations or scientific papers, the way the media presented the Orce Man had an influence on the way that palaeoanthropologists and the so-called scientific community thought about this finding. Likewise, the way that scientists talked about the Orce Man, knowing that their words would be echoed in the media, shows how this channel of communication was also used as a place to define scientific positions and to continue the dispute. In this regard, it is interesting to point to the several instances in which Orce researchers, from both sides of the controversy, sent letters to be published by newspapers. Being the three palaeontology professors' the most significant, to send a letter to the newspapers has been a recurrent way that scientist have

FIGURE 23 A letter sent to Gibert written upon a newspaper report about the Orce Man in which de Lumley again showed his opposition to Gibert's claims
FROM AJG-ICPMC

used to post their opinions on the media.[29] Moreover, several times the media was not only one of the channels used but a necessary channel to establish certain scientific positions. This use of the general media for scientific debate has the effect of multiplying the publics of palaeoanthropology. A painter with no scientific training gave his opinion to Gibert, but politicians also acted and reacted according to what the media said about the case. Funding and permits were given or taken away depending on this media coverage of the Orce Man discovery and controversy. This turned the media into an apparently valid indicator of which scientific position deserved funding and permits. The press was then a source used to decide on the credibility of scientists and their research. The interest that the actors involved had in what the media said is further confirmation of its significance in the story. The Orce Man was a public scientific controversy, a controversy in which the media not only acted as a channel for scientific news or as another popularisation medium, but also as a place where real scientific knowledge was developed, negotiated, and finally recognised (or rejected).

29 At least: De Renzi/Porta/Truyols, *Avui* 1985; Personal Archive Jordi Agustí: Letter from Josep Gibert to the director of *Muy Interesante*, 3 June 1986; Campillo, *El País* 1987; Personal Archive Jordi Agustí: Letter from Moyà-Solà to unknown, 24 september 1987; Gibert, *El Ideal de Granada* 1992; Gibert, *La Vanguardia* 1996; Moyà-Solà/Köhler, *La Vanguardia* 1996; Ribot, *ABC* 1996; García-Olivares, *ABC* 1996; Palmqvist, *El Ideal de Granada* 1996; Gibert, *La Vanguardia* 2005; Ribot, *El Ideal de Granada* 2007. More on the press following in the Annex 2.

Annexes

ANNEX 1

Anatomical Features of the Orce Man

VM-0, the famous Orce Man, is a cranial fragment that consists of two fragments of the parietal bone and a fragment of the occipital bone. It is about 8 centimetres in diameter. At the time of first publication and public presentation, the inner part of the fragment was covered by a piece of rock (figure 3). Despite this, it had, for Gibert, Agustí, and Moyà-Solà, enough characteristics to attribute it to the genus *Homo*: a wide transversal and longitudinal curvature (which indicates a large cranium), thinness (which is typical in infant *Homo*), and the absence of the coronal suture (typical in non-human mammals and not present in the external part).[1] After the cleaning of the inner part, a small crest appeared that was not typical of the *Homo* genus and the possibility of an *Equus* classification was raised (figure 7). In 2002, Bienvenido Martínez-Navarro published an article classifying VM-0 as a female ruminant. For Martínez-Navarro, what were identified as parietal bones, were frontal bones; and what was considered occipital bones, were parietal. During all this time, a series of features were used to classify the fragment as belonging to one genus or another (figure 24).

- *A small internal crest on the occipital bone:* this crest caused the controversy to break out. For the *Equus* proponents, the crest is the norm in this group and in *Homo* it is an exception.[2] Instead, for the *Homo* defenders, this crest lays within *Homo* variability, which is very high for the occipital bone.[3] Gibert and his collaborators detected this crest in an African *Homo ergaster* as did Campillo in a girl from Roman times.[4] For Martínez Navarro large ruminants 'usually' present the crest.[5]
- *Posthumous fracture vs. coronal suture:* Moyà-Solà and Agustí detected a coronal suture in the inner part of VM-0, typical of equines. According to them, this suture is not observed in the external part because it had suffered abrasion.[6] According to Gibert and Campillo, this suture does not exist and what the others see is a posthumous fracture.[7] Campillo even claimed that Moyà-Solà changed the sketch of the supposed suture from one of their articles to another.[8]

1 Gibert/Agustí/Moya-Sola 1983a. See also: Campillo 2002, 171–174.
2 Moyà-Solà/Agustí 1987, 537.
3 Gibert et al. 1989a, 127.
4 Martínez-Navarro 1993, 27–28; Campillo et al. 2006; See also: Martínez-Navarro 1996.
5 Martínez-Navarro 2002, 268. See also Martínez-Navarro 2012.
6 Moyà-Solà/Agustí 1987, 538; Moyà-Solà/Köhler 1997, 96.
7 Gibert et al. 1998a, 212.
8 Campillo 2002, 173.

 | DOI:10.1163/9789004431508_010

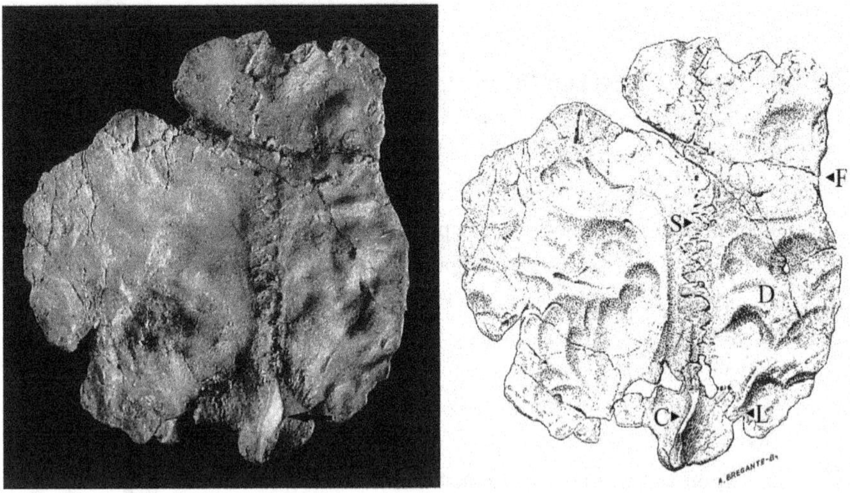

FIGURE 24 Photo (left) and drawing (right) of VM-0. Left, the arrow points to the crest that first caused the controversy. Right, C indicates this crest, S is the sagittal suture, L is the lambdoid suture, F is the posthumous fracture or the coronal suture depending on the interpretations, and D shows four clear digital markings
FROM MARTÍNEZ-NAVARRO 1996

Martínez-Navarro also considered this feature as a fracture and stated that Moyà-Solà and Agustí were wrong in consider it as a suture.[9]

– *Sagittal and lambdoid sutures*: In their early articles neither Gibert nor Agustí and Moyà-Solà noted the complexity of the sagittal and lambdoid sutures as an element that could include or exclude VM-0 from hominid classification.[10] Yet, later, Gibert and Palmqvist used fractal analysis of these sutures to conclude that their shape fit within the range of hominids and humans.[11] Two years later, Palmqvist performed the same analysis with a different drawing of VM-0, which led to a different conclusion (see figure 15).[12] For Martínez-Navarro, the sagittal sutures were in fact interfrontal sutures between both frontal bones, and lambdoid sutures, were coronal sutures; which matched, according to Martínez-Navarro, the complexity of the Orce Man bone fragment.[13]

9 Martínez-Navarro 2002, 266.
10 Gibert/Agustí/Moyà-Solà 1983a; Gibert 1984, and Moyà-Solà/Agustí 1987.
11 Gibert/Palmqvist 1995.
12 Palmqvist 1997.
13 Martínez-Navarro 2002, 267.

- *Digital markings*: Both sides agreed that the inner part of the fragment presents small and shallow digital markings.[14] According to those that were against the Orce Man, the *Homo* genus 'usually' does not present digital impressions yet equines do.[15] On the other hand, Campillo stated that these markings are normal in growing craniums.[16] For Martínez-Navarro, "most ruminants possess deep digital impressions digital in the frontal bone".[17]
- *Longitudinal and transversal curvature*: As already stated, the curvature of the cranium indicates a large brain, typical of the genus *Homo*.[18] Moyà-Solà and Agustí acknowledged this fact but argued that young horses and rhinoceros could also have this curvature.[19] Agustí also stated that the curvature was caused by posthumous modification due to sediment weight.[20]

---

14 The inner depressions on the cranium that correspond to the ridges (sulci and gyri) of the brain are called digital markings. These are marked when the cranium is in formation. Liem 2002, 53.

15 Moyà-Solà/Köhler 1997, 94–95.

16 Campillo 2002, 172.

17 Martínez-Navarro 2002, 268.

18 Campillo 2002, 174.

19 Moyà-Solà/Köhler 1997, 92. Interview with Agustí 2012.

20 Interview with Agustí 2012.

ANNEX 2

News from 1983 to 1999

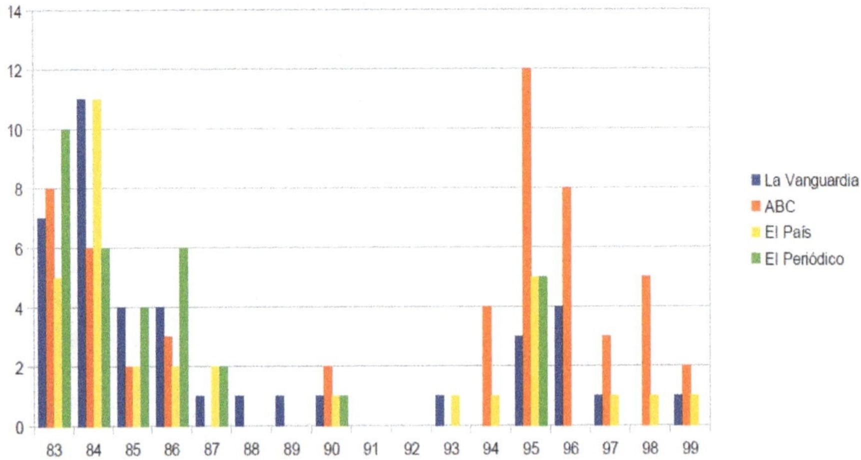

FIGURE 25 News pieces dealing with Orce from 1983 to 1999 in four different Spanish newspapers

This graphic is a representation of the numbers of news items in selected newspapers devoted to the Orce controversy from 1983 to 1999 (figure 25). It is not a comprehensive and detailed compilation of the items. The articles selected are significant because of its extension, relevance in the story and use in this book. Brief notes, side news and non-relevant pieces have been excluded. The graphic does not go beyond 1999 because in the early 20th-century the media situation changed with the expansion of the on-line versions of these and other newspapers.

The resulting graphic shows, as expected, that the years of the discovery of the Orce Man, the subsequent explosion of the controversy, in 1983–1984, and the 1995 Orce Conference with its aftermaths, represent the peaks of the press attention to the Orce controversy. The late 1980s and the early 1990s are the periods where the interest on Orce is weaker.

The representation also shows how the reporting on Orce of different newspapers changed in these different phases of the controversy. The attention to the case during the first years was spread in all the Spanish and Catalan main newspapers, *La Vanguardia, ABC, El País, El Periódico. ABC, El Periódico* and maybe also *La Vanguardia* showed more interest on the discovery, while *El País*, probably because it had the scoop (figure 8), published much more articles on the controversy that on the previous discovery. *El Periódico* was the only paper who actively followed the controversy in the period from 1985–1987. Similarly, *ABC*, maybe because of its Andalusian office, had the lead on the interest on the Conference and the next controversial years.

 | DOI:10.1163/9789004431508_011

ANNEX 3

The Travelling Bone

In order to highlight the relevance of the materiality of the bone as an analytical category to understand the story of the Orce Man controversy, this table shows the travelling story of the Orce fragment since it was discovered, in July 1982, to the end of 1984, when it was kept for a long time in Sabadell under Gibert's control (figure 26). This representation shows, with the colours, the location of the bone and also if the bone was in a private environment or it was shown public, with the black or red spots.

During these first crucial years, the bone remained mostly only accessible to very few people in Sabadell, where Salvador Moyà-Solà first separated it from a big rock. Later, the bone was also for a long time also in the Barcelona Archaeological Museum, where the inner part was cleaned. In both places, the Orce fragment went through scientific preparation before it got its definitive form. Between these two preparations, the bone was taken to the politicians in charge, was shown to relevant international scientists, and, crucially, was publicly presented. Regarding the relevant scientist that saw the bone, in this period, beyond the discoverers, the bone was shown to at least

FIGURE 26 A graphic of both the places where the Orce Man fragment was during the first two crucial years and whether it was shown in public or kept in private

 | DOI:10.1163/9789004431508_012

Crusafont, Mein, Andrews and Campillo before it was public. After that, as long as I know, only the de Lumley couple saw it privately in Orce, and after that, twice in Marseille and once in Paris.

During this first crucial years, the only place where the bone was only publicly exhibited to the general public was Orce, during the excavation campaigns. The first people allowed to see the bone in detail were, therefore, the Orce inhabitants and visitors. The other public appearances of the bone were mostly devoted to the press.

Finally, after these first two years, it remains clear that the bone was under Gibert's control almost from the beginning. Apparently, already just after the first press conference in Granada, Gibert himself took the remain to Orce, and then back to Sabadell, where, for the first time, was kept in a safe in a bank. After that, Gibert himself took the bone to Marseille twice, he was in all press conferences devoted to the bone and he took the bone to Perpignan.

Bibliography

1 Interviews

All interviews were conducted by the author, except indicated.

Interview with Josep Gibert, 2003, conducted by Carlos Acosta Rizo. *Institut de Paleontologia*, Sabadell (Barcelona).

Interview with Domènec Campillo, 5 June 2011. Museu d'Arqueologia de Catalunya, Barcelona.

Interview with Jordi Agustí, 9 July 2012. Conducted by the author with Oliver Hochadel, Institució Milà i Fontanals, CSIC, Barcelona.

Interview with Lluís Gibert, 9 and 27 November 2012. Centre Cívic Can Deu, Barcelona.

Interview with Salvador Moyà-Sola, 28 November 2012. Universitat Autònoma de Barcelona, Bellaterra (Barcelona).

Interview with Enrique García-Olivares, 28 January 2013 and 28 October 2014. By e-mail.

Interview with Michael J. Walker, 5 February 2013. By e-mail.

Interview with Roger Marcet, 19 December 2013. Museu Martítim de Barcelona, Barcelona.

Interview with Nora Moloney, 11 March 2014. By e-mail.

2 Archives

Arxiu Històric de la Diputació de Barcelona (ADB), Barcelona.

Arxiu Miquel Crusafont (AMC-ICPMC), Institut Català de Paleontologia Miquel Crusafont, Sabadell.

Arxiu Josep Gibert (AJG-ICPMC), Institut Català de Paleontologia Miquel Crusafont, Sabadell.

Personal Archive Jordi Agustí.

Personal Archive Lluís Gibert, including Orce Conference Video Recording (OCVR), 1995.

3 Secondary Literature

Acosta, C. 2013. 'Un pie en el cielo y otro en la ciencia. Las interacciones productivas de Miquel Crusafont (1910-1983)'. *Dynamis* 33, 2: 321–342.

Aczel, A. 2007. *The Jesuit and the skull: Teilhard de Chardin, evolution, and the search for Peking man*. London: Penguin.

Aguirre, E.; Arsuaga, J. L.; Bermúdez de Castro, J. M.; Carbonell, E.; Ceballos, M.; Díez, C.; Enamorado, J.; Fernández-Jalvo, Y.; Gil, E.; Gracia, A.; Martín-Nájera, A.; Martínez, I.; Morales, J.; Ortega, A. I.; Rosas, A.; Sánchez, A.; Sánchez, B.; Sesé, C.; Soto, E.; Torres, T. J. 1990. 'The Atapuerca sites and the ibeas hominids'. *Human Evolution* 5, 1: 55–73.

Agustí, J.; Gibert, J.; Moyà-Solà, S. 1983b. 'L'Home d'Orce: el primer europeu'. *Arrel* 6: 71–79.

Agustí, J.; Gibert, J.; Moyà-Solà, S. 1983c.'El Hombre de Orce'. *Revista de Arqueología* 29: 16–21.

Agustí, J.; Moyà-Solà, S. 1987. 'Sobre la identidad del fragmento craneal atribuido a *Homo* sp. en Venta Micena (Orce, Granada)'. *Estudios Geológicos* 43, 5–6: 535–538.

Agustí, J.; Lordkipanidze, D. 2005. *Del Turkana al Cáucaso. La evolución de los primeros pobladores de Europa* Barcelona: Adventure Press National Geographic.

Allan, S. (ed.). 2011. 'Introduction: Science Journalism in a Digital Age'. *Journalism* 12, 7: 771–777.

Andrews, P. 2011. 'Curriculum Vitae' in: https://create-cdn.net/sitefiles/29/7/8/297891/Dr_Peter_Andrews.pdf [Accessed January 2020].

Anonymous. 1983a. 'L'Institut de Paleontologia: Una institució científica poc coneguda'. *Arrel* 6: 63–66.

Anonymous. 1983b. 'El jaciment paleontològic de Venta Micena' *Arrel* 6: 67–70.

Anonymous. 1983c. 'Hombre de Orce' *Historia y Vida* 187: 32–42.

Anonymous. 2009. 'Yves Coppens: 50 ans au service de la paléoanthropologie', 13 February, in: Hominidés.com, http://www.hominides.com/html/biographies/yves_coppens.php [Accessed January 2020].

Anonymous. 'Xavier Folch i Recasens', in: *Gran Enciclopèdia Catalana*, http://www.enciclopedia.cat/EC-GEC-0519403.xml [Accessed January 2020].

Anonymous. 'Un peu d'histoire …', in: *Musée de Tautavel*, http://www.450000ans.com/ [Accessed January 2020].

Anonymous. 'François Sémah', in: *Futura-Sciences.com*, http://www.futura-sciences.com/magazines/sciences/infos/personnalites/d/homme-francois-semah-529/ [Accessed January 2020].

Anonymous. 'Ian Tattersall', in: *American Museum of Natural History*, http://www.amnh.org/our-research/staff-directory/ian-tattersall [Accessed January 2020].

Antena 3 1995. Noticias 2, 5 September. [Video Recording].

Arribas, A.; Palmqvist, P. 2002. 'The first human dispersal to Europe: Remarks on the archaeological and palaeoanthropological record from Orce (Guadix-Baza basin, southeastern Spain)'. *Human Evolution* 17, 1–2: 55–77.

Arsuaga, J.L.; Martínez, I. 1998. *La especie elegida. La larga marcha de la evolución humana*. Gavà (Barcelona): Temas de Hoy.

Ascenzi, A.; Biddittu, I.; Cassoli, P.F.; Segre, A.G.; Segre-Naldini, E. 1996. 'A calvarium of late *Homo erectus* from Ceprano, Italy'. *Journal of Human Evolution* 31, 5: 409–423.

Balart, G. 2007. *Humans a Venta Micena* Doble Banda productions [Documentary].

Baltas, A. 2000. 'Classifying scientific controversies' in: Machammer, P.; Pera, M.; Baltas, A. (eds.) *Scientific controversies. Philosophical and Historical Perspectives.* Oxford: Oxford University Press, 40–49.

Balter, M. 2009. 'Early Europeans Used Sophisticated Tools', 2 September. *Science News*, http://news.sciencemag.org/europe/2009/09/early-europeans-used-sophisticated-tools [Accessed January 2020].

Beauchamp, T. 1987. 'Ethical Theory and the problem of closure' in: Engelhardt, H. T.; Caplan, A. L. (eds.) *Scientific Controversies* New York: Cambridge University Press, 27–48.

Bednarik, R. 1989. 'On the Pleistocene settlement of South America'. *Antiquity* 63: 101–111.

Bermúdez de Castro, J. M.; Nicolás, M. E. 1996. 'Changes in the lower premolar-size sequence during hominid evolution. Physlogenetic implications'. *Human Evolution* 11, 3–4: 205–215.

Bermúdez de Castro, J. M. 2013 'La colonización de Europa', in: *Blog de José María Bermúdez de Castro. Reflexiones sobre la evolución humana*, http://josemariabermudezdecastro.com/prueba/?p=69#comments [Accessed January 2020].

Boëda, E. et al. 2014a. 'A new late Pleistocene archaeological sequence in South America: the Vale da Pedra Furada (Piauí, Brazil)'. *Antiquity* 88, 341: 927–941.

Boëda, E. et al. 2014b. 'The peopling of South America: expanding the evidence'. *Antiquity* 88, 341: 954–955.

Bonifay, E. 1981. 'Les traces des priemiers hominides en France'. *La Recherche* 128, 12: 1442–1444.

Bonifay, E.; Vandermeersch, B. (eds.). 1991. *Les premiers européens. Actes du 114ª Congres National des Société Savantes (Paris, 3-9 avril 1989)* Paris: Editions du C.T.H.S.

Bonifay, E. 2002. *Les premiers peuplements de l'Europe* Paris: La maison des roches.

Borja, C.; García-Pacheco, J.M.; Ramírez-López, J.P.; García-Olivares, E. 1992. 'Cuantificación y caracterización de la albúmina fosil del craneo de Orce' in: Gibert, J.; García-Olivares, E.; Malgosa, A.; Martínez-Navarro, B. (eds.). *Presencia humana en el Pleistoceno Inferior de Granada y Murcia* Orce, Granada: Museo de Prehistoria, 415–423.

Borja, C. 1995. *Detección y caracterización de proteínas fósiles mediante técnicas inmunes.* Doctoral thesis, Facultad de Medicina, Universidad de Granada.

Borja, C.; García-Pacheco, M.; García-Olivares, E.; Scheuensthul, G.; Lowenstein, J. 1997. 'Immunospecificity of albumin detected in 1.6 million-year-old fossils from Venta Micena in Orce, Granada, Spain'. *American Journal of Physical Athropology* 103: 433–441.

Borrero, L. A. 1995. 'Human and natural agency: some comments on Pedra Furada'. *Antiquity* 69: 602–3.

Bowman-Kruhm, M. 2005. *The Leakeys: a biography* Westport: Greenwood biographies.

Bower, B. 1997. 'Ancient roads to Europe: African ancestors may have entered Europe surprisingly early'. *Science News* 151, 1: 12–13.

Bower, B. 2009. 'Europe's oldest stone hand axes emerge in Spain'. *Science News*, September 2. https://www.sciencenews.org/article/europes-oldest-stone-hand-axes-emerge-spain [Accessed January 2020].

Bray, W. 1986. 'Finding the earliest Americans'. *Nature* 321: 726.

Bray, W. 1988. 'The palaeoindian debate'. *Nature* 332: 107.

Brossard, D. 2013. 'New media landscapes and the science information consumer'. *Proceedings of the National Academy of Sciences of United States of America*, 110 (3): 14096–14101.

Bubela, T. et al. 2010. 'Science communication reconsidered'. *Nature Biotechnology* 27, 6: 514–518.

Bucchi, M. 1996. 'When scientists turn to the public: Alternative routes in science communication'. *Public Understanding of Science* 5: 375–394.

Bucchi, M. 1998. *Science and the Media* London-New York: Routledge.

Bucchi, M. 2017. 'Credibility, expertise and the challenges of science communication 2.0' *Public Understanding of Science* 26, 8:890–893.

Büchi, M. 2017. 'Microblogging as an extension of science reporting' *Public Understanding of Science* 26, 8: 953–968.

Burbuja, Foro de Economía. 2013. 'Noticia: Andalucia, no tenemos pa comer y se lo gastan en buscar un diente' 2 March, in: *Burbuja, Foro de Economía.* http://www.burbuja.info/inmobiliaria/burbuja-inmobiliaria/399561-andalucia-no-pa-comer-y-se-gastan-buscar-diente.html [Accessed January 2020].

Calvo, J. 2015. *Altamira.* Barcelona: Stella Maris.

Camprubí, L. 2014. *Engineers and the Making of the Francoist Regime.* Cambridge MA: MIT Press.

Campillo, D. 2002. *El cráneo infantil de Orce: el homínido más antiguo de Eurasia.* Barcelona: Bellaterra.

Campillo, D.; Cuesta, MM; García-Guixé, E.; Chimenos, E.; Devenat, L.; Baxarias, J. 2006. 'An occipital crest in an infant cranium from the Roman necropolis of Francolí (Tarragona, Spain): implications to the interpretation of the Orce skull'. *Revista Española de Antropología Física* 26: 93–101.

Campillo, D. 2010. 'El Hombre de Orce es un homínido'. *Revista de Arqueología* 35: 14–21.

Canal Plus 1995. Redacción, 7 September. [Video Recording].

Canal Sur 1995a. El Diario 1, 1 September. [Video Recording].

Canal Sur 1995b. El Diario 2, 3 September. [Video Recording].

Canal Sur 1995c. El Diario 1, 4 September. [Video Recording].

Canal Sur 1995d. El Diario 2, 4 September. [Video Recording].

Canal Sur 1995e. El Diario 2, 6 September. [Video Recording].

Canal Sur 1995f. El Diario 1, 8 September. [Video Recording].

Canal Sur 1995g. Los Reporteros, July 1995. [Video Recording].

Canal Sur 1997. Los Reporteros, May 1997. [Video Recording].

Carandell, M. 2012. 'A Different Dimension: Center-Periphery relations due to the Orce Man Public Controversy' In: *8th STEP Meeting, Corfú 2012*.

Carandell, M. 2013a. 'Newspapers as the arena of scientific controversy: The debate about the Orce man in Spanish mass-media'. In: Allibone, L.; Locke, S. (eds.) *Knowledges in Publics: Beyond Deficit, Engagement and Transfer* Newcastle upon Tyne: Cambridge Scholars Publishing, 150–170.

Carandell, M. 2013b. 'Hominidos, dudas y grandes titulares: la controversia del Hombre de Orce en la prensa española (1983–2007)'. *Dynamis* 33, 2: 365–387.

Carandell, M. 2015a. 'Hemos venido, hemos visto y se nos ha conquistado: Los congresos cientificos como instrumentos de validación en controversias' in: Bellver Loizaga, V.; d'Amaro, F.; Molina Puertos I.; Ramos Tolosa, J. (eds.) *"Otras voces, otros ámbitos": Los sujetos y su entorno. Nuevas perspectivas de la historia sociocultural. Ciencia, medicina y sociedad: debates y perspectivas en historia de la ciencia*. València: Asociación de Historia Contemporánea-Universitat de València.

Carandell, M. 2015b. *Orce Man. A public controversy in Spanish human origins research 1982–2007*. PhD thesis, Universitat Autònoma de Barcelona.

Carandell, M. 2016. 'The First American Scoop: The Pedra Furada Controversy in Newspapers (1978-2015)' in Hochadel, O.; Carandell, M.; Florensa, C. 2016. 'Scuffles, Scoops and Scams. The Construction of Prehistoric Knowledge in Newspapers'. *Centaurus* 58, 3, 239–256.

Carbonell, E.; Estevez, J.; Moyà-Solà, S.; Pons-Moyà, J.; Agusti, J.; Villalta, J.F. 1981. 'Cueva Victoria (Murcia, España): lugar de ocupaci6n humanamas antiguo de la peninsula ibérica'. *Endins* 8: 47–57.

Carbonell, E.; Estévez, J. 1983. 'Les primeres occupacions humanes en els Països Catalans'. *Revista Avenç* 57: 151–154, (67–70).

Carbonell, E.; Bermúdez de Castro, J. M.; Arsuaga, J. L.; Díez, J. C.; Rosas, A.; Sala, R.; Mosquera, M.; Rodríguez, X. P. 1995. 'Lower Pleistocene Hominids and Artifacts from Atapuerca-TD-6 (Spain)'. *Science* 269: 826–830.

Carbonell, E.; Sala, R. 2000. *Planeta Humà* Barcelona: Editorial Empúries.

Carbonell, E.; Sala, R. 2002. *Aún no somos humanos* Barcelona: Ediciones Península.

Cardeñosa, B. 2007 'El Hombre de Orce el primer habitante de la península'. *Historia de iberia vieja* 29: 78–81.

Cardús, S. 1999. 'The Circulation of Daily Newspapers in the Catalan-speaking areas between 1976 and 1996'. *The Annual Joan Gili Memorial Lecture*. Sheffield: The Anglo-Catalan Society.

Català Gorgues, J. 2013. 'Miquel Crusafont, George Simpson y la internacionalización de los estudios de paleontología evolutiva en España'. *Dynamis* 33, 2: 343–364.

Cattaneo, C.; Gelsthorpe, K.; Phillips, P.; Sokol, R. J. 1993. 'Blood residues on stone tools: indoor and outdoor experiments'. *World Archaeology* 25: 29–43.

Cela-Conde, C. 1996. '¿Comienza Africa en los pirineos?' *Investigacion y Ciencia.* 232: 34–36.

Chadarevian, S. 1997. 'Using interviews to write the history of science' in: Söderqvist, T. (ed.) *The historiography of contemporary science and technology, Studies in the history of science, technology & medicine*, Amsterdam: Harwood Academic Publishers, 51–70.

Chamorro, E. 1983. 'El Peregrino de Orce'. *Cambio 16* 605: 84–87.

Champion, T.; Gamble, C.; Shennan, S.; Whittle, A. 1984. *Prehistoric Europe* London: Academic Press.

Chew, C.; Eysenbach, G. 2010. 'Pandemics in the Age of Twitter: Content Analysis of Tweets during the 2009 H1N1 Outbreak'. *PLoS ONE* 5, 11: e14118.

Chippindale, C. 1986. 'At home with the First European'. *New Scientist* 9: 38–41.

Clemens, E. 1986. 'Of asteroids and dinosaurs: The role of the press in the shaping of scientific debate'. *Social Studies of Science* 16, 3: 421–456.

Cody, L. 2010. *Engaging with Online Audiences: A cross-country analysis of use of the Web for communicating biomedical science to public audiences.* Master of Arts thesis, Dublin City University.

Cohnen, F.; Gómez, J. E. 1995. 'Paleontólogos españoles se disputan el hallazgo del primer europeo'. *Tribuna,* 4 de Septiembre, 58–61.

Cole, S. 1975. *Leakey's luck: The life of Louis Leakey, 1903-72.* London: Collins.

Collins, H. M. 1981. 'Son of Seven Sexes: The Social Destruction of a Physical Phenomenon'. *Social Studies of Science* 11: 33–62.

Collins, H. M. 1985. *Changing order: Replication and induction in scientific practice* Chicago: The University of Chicago Press.

Collins, H. M. 2017. *Gravity's Kiss: The Detection of Gravitational Waves.* Cambridge, MA: MIT Press.

Colson, V. 2011. 'Science blogs as competing channels for the dissemination of science news'. *Journalism* 12, 7: 889–902.

Coltorti, M. et al. 1982. 'Reversed magnetic polarity at an early Lower Palaeolithic site in Central Italy'. *Nature* 300: 173–176.

Contreras, A. L. 2006. 'La Dama de Baza: hemerografía.' *Lucentum* 25: 125–137.

Coppens, Y. 2005. (1997). *La Rodilla de Lucy* Barcelona: Tusquets Editores.

Cuatro, Cuarto Milenio 2013. 'Los 200 del Everest' 17 March 2013, Temporada 8 Programa 312. [Video Recording].

Dart, R. 1959. *Adventures with the missing link* Britain: Hamish Hamilton, Ltd.

De Bont, R. 2003. 'The Creation of Prehistoric Man'. *Isis* 94, 4: 604–630.

Delisle, R. G. (ed.) 2012. 'Human Evolution Across Disciplines: Through the Looking Glass of History and Epistemology'. *History and Philosophy of the Life Sciences*, Special issue 34, 1–2.

Denison, S. 1995. '1,8 million-year-old human presence claimed in Spain' *British Archaeology* 7: 4.

Dennell, R.; Roebroeks, W. 1996. 'The earliest colonization of Europe: the short chronology revisited'. *Antiquity* 70: 535–542.

Díaz-Andreu, M. 2012. *Archaeological Encounters: Building Networks of Spanish and British Archaeologists in the 20th Century* Newcastle upon Tyne: Cambridge Scholars Publishing.

Díaz Rojo, J. A. 2007a. 'La modalidad epistémica asertiva como recurso retórico en noticias científicas. el caso del hallazgo paleontológico del "Hombre" de Orce' *Revista de lingüística y lenguas aplicadas* 2: 15–26.

Díaz Rojo, J. A. 2007b. 'Paleoantropologia i premsa: el tractament periodístic de l'Home d'Orce.' *Mètode: Revista de difusió de la investigació de la Universitat de Valencia* 53: 37–42.

Díaz Rojo, J. A. 2011. 'Retórica científica en la prensa. El hallazgo paleontológico del cráneo de Orce (1983).' in: Díaz Rojo, J. A. (ed.) *La circulación del saber científico en los siglos XIX y XX,* València: Instituto de Historia de la Medicina y de la Ciencia López-Piñero, 99–128.

Diputació de Barcelona. 1985. *Origen i Evolució de l'Home* Santa Perpetua de la Moguda (Barcelona): Gabinet de Publicacions de la Diputació de Barcelona.

Domènech i Sampere, X. 2002. 'El cambio político (1962-1976): materiales para una perspectiva desde abajo', *Historia del presente*, nº1, pp. 46-67.

Drévillon, E. 2011. *Le Secret de la Roche Percée* Mayenne, France: Fayard.

Durán, F. 2012 '1994-1996: la pinza' 20 May, in: *Crónicas de un Escéptico* http://blogs.grupojoly.com/esceptico/2012/05/20/1994–1996-la-pinza/ [Accessed January 2020].

Engelhardt, H. T.; Caplan, A. L. (eds.) 1987. *Scientific Controversies* New York: Cambridge University Press.

Facebook Cuarto Milenio 2013. 'Retirado "temporalmente" el estudio sobre el primer europeo conocido.' 21 March, in: *Facebook Cuarto Milenio.* https://m.facebook.com/cuartomileniotv/photos/a.398810520213878.96010.148134 965281436/431013666993563/?type=1&p=0 [Accessed January 2020].

Fahnestock, J. 1989. 'Arguing in different forums: The Bering crossover controversy'. *Science, Technology & Human Values* 14, 1: 26–42.

Fiedel, S. J. 1996. 'Blood from stones? Some methodological and interpretive problems in blood residue analysis'. *Journal of Archaeological Science* 23: 139–147.

Florensa, C. 2013. 'Breaking the silence: Palaeontology and evolution in La Vanguardia Española (1939-1975).' *Dynamis* 33, 2: 297–320.

Florensa, C. 2017. *Els discursos sobre l'evolució en el franquisme (1939-1967). La Generación del 48 i 'La evolución sin problema'* PhD Thesis. Universitat Autònoma de Barcelona.

Funari, P. P. 2000. 'Archaeology, education and Brazilian identity'. *Antiquity* 74: 182–5.

Furió, M. 2013. 'Un diente de homínido en Orce' 21 March, 2013, in: *Los Fósiles Hablan,* http://www.investigacionyciencia.es/blogs/medicina-y-biologia/15/posts/un-diente-de-homnido-en-orce-10972 [Accessed January 2020].

Gamble, C. 1994. 'Time for Boxgrove man'. *Nature* 369: 275–276.

García Naharro, F. 2011. 'Diálogo de discursos en el período de la transición a la democracia' *Anales del Instituto de Estudios Madrileños*, nº51, pp. 437-453.

García-Olivares E.; Gallardo, J.M.; Martinez, F.; Borja, C.; García-Olivares, D. 1989. 'Detección y caracterización de proteínas fósiles en el cráneo de Orce (resultados preliminares)' in: Gibert, J.; Campillo, D.; García-Olivares, E. (eds.) *Los restos humanos de Orce y Venta Micena,* Sabadell: Institut Paleontologic Miquel Crusafont, 225–228.

Garcia Ruiz, R. 2003. 'L'Origen de l'Home, els fills de la Terra', unpublished High School project, Cornellà de Llobregat: IES Miquel Martí i Pol. Supervisor: Agustí Camòs.

Gaspar Neto, V.; Santos, R. V. 2009. 'The color of the bones: scientific narratives and cultural appropriations of "Luzia", a prehistoric skull from Brazil'. *Mana* 15, 2: 449–480.

Gavroglu, K.; Patiniotis, M.; Papanelopoulou, F.; Simoes, A.; Carneiro, A.; Diogo, M.P.; Bertomeu, J.R.; García-Belmar, A.; Nieto-Galan, A. 2008. 'Science and technology in the European periphery: Some historiographical reflections'. *History of Science* 46: 153–175.

Gibbons, A. 2002. 'Glasnost for Hominids: Seeking access to fossils'. *Science* 297, 5586: 1464–1468.

Gibbons, A. 2007. *The First Human. The race to discover our earliest ancestors* New York: Anchor Books.

Gibert, J.; Agustí, J.; Moya-Sola, S. 1983a. 'Presencia de *Homo* sp. en el yacimiento del Pleistoceno inferior de Venta Micena (Orce, Granada)' *Paleontologia i Evolucio* special issue 1983: 1–9.

Gibert, J. 1984. 'El yacimiento de Venta Micena (Orce, Granada). Su importancia, acción antropogénica y características paleoantropologicas del fragmento de cràneo de *Homo* sp'. in: Arteaga, O. (ed.) *Homenaje a Luis Siret (1934 a 1984). Actas del Congreso Homenaje a Luis Siret (1934-1984), Cuevas de Almanzora, Junio, 1984,* Madrid, Dirección General de Bellas Artes de la Consejería de Cultura de la Junta de Andalucia, 37–49.

Gibert, J.; Campillo, D.; García-Olivares, E.; Gallardo, J.M.; Martinez, F.; Borja, C.; García-Olivares, D. 1986. 'Description and differential diagnosis of the fragment of the cranium of *Homo* sp. of Venta Micena locality (Lower Pleistocene).' *The Pleistocene Perspective I, The World Archaelogical Congress*. Southampton.

Gibert, J.; Campillo, D.; García-Olivares, E. 1987. 'Study of the Orce Man'. *2nd International Congress of Human Paleontology, Turin.*

Gibert, J.; Campillo, D.; García-Olivares, E. (eds.) 1989. *Los restos humanos de Orce y Cueva Victoria* Sabadell: Institut Paleontologic Miquel Crusafont.

Gibert, J.; Ribot, F.; Ferrández, C.; Martínez-Navarro, B.; Caporci, R. 1989a. 'Caracteristicas diferenciales entre el fragmento de craneo de *Homo* sp. de Venta Micena (Orce, Granada) y los Equidos'. *Estudios Geológicos* 45: 121–138.

Gibert J.; Ribot F.; Ferrández C.; Martínez-Navarro B.; Caporicci R.; Campillo, D. 1989b. 'Anatomical study: comparison of the cranial fragment from Venta Micena (Orce, Spain) with fossil and extant mammals'. *Human Evolution* 4: 283–305.

Gibert, J.; Jiménez, C. 1991. 'Investigations into cut-marks on fossil bones of Lower Pleistocene age from Venta Micena (Orce, Granada province, Spain).' *Human Evolution*, 4, 1: 7–128.

Gibert, J.; Campillo, D.; Martínez, B.; Sánchez, F.; Caporicci, R.; Jiménez, C.; Ferrández, C.; Ribot, F. 1991. 'Nouveaux restes d'hominid6s dans les gisements d'Orce et de Cueva Victoria (Espagne)' in: Bonifay, E.; Vandermeersch, B. (eds.). *Les premiers européens. Actes du 114ª Congres National des Société Savantes (Paris, 3-9 avril 1989)*. Paris: Editions du C.T.H.S., 273–282.

Gibert, J. et al. (eds.) 1993. *Proyecto Orce-Cueva Victoria (1988-1992). Presencia humana en el Pleistoceno inferior de Granada y Murcia*. Orce: Ayuntamiento de Orce, Museo de Prehistoria José Gibert.

Gibert, J.; Palmqvist, P. 1995. 'Fractal analysis of the Orce skull sutures'. *Journal of Human Evolution* 28: 561–575.

Gibert, J.; Campillo, D.: Arqués, J. M.; García-Olivares, E.; Borja, C.; Lowenstein, G. 1998a. 'Hominid status of the Orce cranial fragment: reasserted'. *Journal of Human Evolution* 34: 203–217.

Gibert, J.; Gibert, Ll.; Iglesias, A.; Maestro, E. 1998b. 'Two Oldowan assemblages in the Plio-Pleistocene deposits of the Orce región, southeast Spain' *Antiquity* 72: 17–25.

Gibert, J. 1999a. 'Del Congreso a las actas: un largo y definitivo camino' in: Gibert, J. et al. (eds.). *The Hominids and their Environment during the Lower and Middle Pleistocene of Eurasia, Proceedings of the International Conference of Human Palaeontology, Orce 1995. Los homínidos y su entorno en el Pleistoceno inferior y medio de Eurasia, Actas del Congreso Internacional de Paleontología Humana, Orce 1995* Orce: Ayuntamiento de Orce, Museo de Prehistoria y Paleontología 'J. Gibert', 9–15.

Gibert, J. et al. (eds.) 1999b. *The Hominids and their Environment during the Lower and Middle Pleistocene of Eurasia, Proceedings of the International Conference of Human Palaeontology, Orce 1995. Los homínidos y su entorno en el Pleistoceno inferior y medio de Eurasia, Actas del Congreso Internacional de Paleontología Humana, Orce 1995* Orce: Ayuntamiento de Orce, Museo de Prehistoria y Paleontología 'J. Gibert'.

Gibert, J.; Malgosa, A.; Sanchez, F.; Ribot, F.; Walker, M. 1999c. 'Humeral fragments attributable to *Homo* sp. from Lower Pleistocene sites at Venta Micena (Orce, Granada, Spain)' in: Gibert, J., et al. (eds.) *The hominids and their enviroment in the middle and lower pleistocene of Europe and Asia. Proceedings of the International Conference of Human Palaeontology, Orce 1995. Los homínidos y su entorno en el Pleistoceno inferior y medio de Eurasia, Actas del Congreso Internacional de Paleontología Humana, Orce 1995* Orce: Ayuntamiento de Orce, Museo de Prehistoria y Paleontología 'J. Gibert', 87–112.

Gibert, J.; Gibert, Ll.; Albadalejo, S.; Ribot, F.; Sánchez, F.; Gibert, P. 1999d. 'Molar tooth fragment BL5-0: the oldest human remain found in the Plio-Pleistocene of Orce (Granada province, Spain)'. *Human Evolution* 14, 1–2: 3–19.

Gibert, J.; Gibert, Ll. 2000. 'Las investigaciones de Orce y los medios de comunicación'. *II Simposio Latino sobre Geología, Medio Ambiente y Sociedad.* March 2000, 28: 227–232.

Gibert, J.; Sánchez, F.; Ribot, F. 2002. 'Human remains in the Lower Pleistocene sediments from the Orce and Cueva Victoria areas (South Eastern Spain)'. *L'Antropologie* 106: 669–683.

Gibert, J.; Gibert, Ll.; Iglesias, A. 2003. 'The Gibraltar Strait: A Pleistocene door of Europe?'. *Human Evolution* 18, 3–4: 147–160.

Gibert, J. 2004. *El Hombre de Orce, los Hominidos que llegaron del Sur* Córdoba: Almuzara.

Gibert, J.; Gibert, L. 2005. 'First people in Spain'. *Current World Archaeology* 20: 22–28.

Gibert, J. 2007. *¿Quieres ser paleontólogo?* Barcelona: EDBE.

Gieryn, T. F. 1983. 'Boundary-work and the demarcation of science from non-science: strains and interests in professional ideologies of scientists'. *American Sociological Review* 48, 6: 781–795.

Gieryn, T. F. 1999. *Cultural boundaries of science: Credibility on the line* London: The University of Chicago Press.

Golinski, J. 1998. *Making Natural Knowledge. Constructivism and the History of Science* Chicago: Cambridge University Press.

Golson, J. 1986. 'The World Archaeological Congress, Southampton, and its aftermath'. *Australian Archaeology* 23: 100–105.

González-Silva, M. 2005. 'Del factor sociológico al factor genético. Genes y enfermedad en las páginas de *El País* (1976-2002)'. *Dynamis* 25: 487–512.

González-Silva, M. 2007. 'Funding through the press: Genomics as a new political issue in the pages of *El País*' in: Herran, N. et al. (eds.) *Synergia: Primer encuentro de jóvenes investigadores en historia de la ciencia* Madrid: CSIC.

González-Silva, M. 2009. 'With or Without Scientists: Reporting on Human Genetics in the Spanish Newspaper *El País* (1976-2006)'. In: Papanelopoulou, Nieto-Galan and Perdiguero (eds.) 2009. *Popularizing Science and Technology in the European Periphery, 1800-2000* London: Ashgate, 217–236.

González-Silva, M. 2014. *Genes de Papel. Genética, Retórica y Periodismo en el diario El País (1976-2006)*. Madrid: CSIC, Estudios sobre la ciencia. Vol. 65.

Goodell, R. 1977. *The Visible Scientists* Boston: Little, Brown and Co.

Goodell, R. 1987 'Role of mass media in scientific controversies' In: Engelhardt, H.; Caplan A.L. (eds.) 1987. *Scientific Controversies* New York: Cambridge University Press, 585–598.

Goodrum, M. 2000. 'The history of twentieth century paleoanthropology: A bibliographic survey'. *History of Anthropology Newsletter* 2: 10–15.

Goodrum, M. 2009. 'The history of human origins research and its place in the history of science: research problems and historiography'. *History of Science* 47, 3: 337–357.

Goodrum, M. 2013. 'Obituary: Phillip Vallentine Tobias (1925-2012)'. *American Journal of Physical Anthropology* 150: 167–169.

Gore, R. 'The Dawn of Humans: The First Europeans'. *National Geographic* July 1997, 96–113.

Goulden, M. 2009a. 'Boundary-work and the human-animal binary: Piltdown man, science and the media'. *Public Understanding of Science* 18: 275–291.

Goulden, M. 2009b. *Excavating humanness: palaeoanthropology at the human-animal boundary* PhD Thesis. Nottingham: Univeristy of Nottingham.

Goulden, M. 2013. 'Hobbits, hunters and hydrology: Images of a "missing link," and its scientific communication'. *Public understanding of science* 22: 575–89.

Gregory, J.; Miller, S. 1998. *Science in Public: Communication, Culture and Credibility* New York: Basic Books.

Gregory, J. 2003. 'The popularisation and excommunication of Fred Hoyle's "life-from-space" theory'. *Public Understanding of Science* 12: 25–46.

Gregory, J. 2005. *Fred Hoyle's Universe* New York: Oxford University Press.

Gross, A. G. 1996. *The rhetoric of science*. Cambridge M.A.: Harvard University Press.

Gruhn, R. 1987. 'On the Settlement of the Americas: South American Evidence for an Expanded Time Frame'. *Current Anthropology* 28, 3: 363–365.

Gruhn, R.; Bryan, A. L. 1991. 'A Review of Lynch's descriptions of South American Pleistocene sites'. *American Antiquity* 56, 2: 342–348.

Guàrdia, C.; Pou, F. 1996. 'Pedres que parlen' 30 minuts TV3, in: *TV3 a la carta*, http://www.tv3.cat/videos/1390199 [Documentary, accessed January 2020].

Gutiérrez, M.; Karlin, C.; Faure, M. 2016 'Claude Guérin (1939-2016)', *Afrique: Archéologie & Arts* online: http://journals.openedition.org/aaa/890 [Accessed January 2020].

Guidon, N. 1983. 'Contribution à l'étude de l'art rupestre de l'Amerique du Sud'. *L'Anthropologie* 87, 2: 257–270.

Guidon, N. 1984. 'Les premières occupations humaines de l'aire archéologique de São Raimundo Nonato-Piauí-Brasil'. *L' Antropologie* 88, 2: 263–271.

Guidon, N.; Delibrias, G. 1986. 'Carbon-14 dates point to man in the Americas 32.000 years ago'. *Nature* 321: 769–771.

Guidon, N. et al. 1996. 'Nature and age of the deposits in Pedra Furada, Brazil: reply to Meltzer, Adovasio & Dillehay'. *Antiquity* 70: 408–415.

Gundling, T. 2010. 'Human Origins Studies: A Historical Perspective'. *Evolution: Education and Outreach* 3: 314–321.

Gusi, F. 2005a. 'Introducción al yacimiento' in: Gusi, F. (ed.) *Roedores, monos, caballos y ciervos: faunas fósiles de Casa Blanca-Almenara de 5 a 1 millones de años* (Castellón)

Diputació de Castelló, Servei d'Investigacions Arqueológiques i Prehistòriques, 17–30.

Gusi, F. (ed.) 2005b. *Roedores, monos, caballos y ciervos: faunas fósiles de Casa Blanca-Almenara de 5 a 1 millones de años (Castellón)* Diputació de Castelló, Servei d'Investigacions Arqueológiques i Prehistòriques.

Hammond, M. 1982. 'The expulsion of the Neanderthals from human ancestry: Marcellin Boule and the social context of scientific research'. *Social Studies of Science* 12, 1: 1–36.

Hawks, J. et al. 2017. 'New fossil remains of Homo naledi from the Lesedi Chamber, South Africa', *eLife*, May 9, 2017.

Hedges, S. B.; Schweitzer, M. H. 1995. 'Detecting dinosaur DNA'. *Science* 268, 5214: 1191–1192.

Hedges, S. B.; Schweitzer, M. H. 1999. 'Molecular palaeontology' in: Singer, R. (ed.) *Encyclopedia of Paleontology.* Chicago: Fitzroy-Dearborn Publishers, 752–754.

Hernández, J. 1984. '¿Quiénes somos? ¿De dónde venimos? ¿A dónde vamos?', in: Siniestro Total, *Menos mal que nos queda Portugal.* Madrid: Dro. [Music Record].

Hevly, B. 1996. 'The heroic science of glacier motion'. *Osiris* 11: 66–86.

Hilgartner, S. 1990. 'The dominant view of popularisation: conceptual problems, political uses'. *Social Studies of Science* 20, 3: 519–539.

Hobsbawm, E.; Ranger, T. 1992. *The Invention of Tradition* Chicago: Cambridge University Press.

Hochadel, O. 2009. 'Atapuerca – the Making of a Magic Mountain. Popular Science Books and Human-Origins- Research in Contemporary Spain'. In: Schirrmacher, Arne (ed.) *Communicating Science in 20th Century Europe. A Survey on Research and Comparative Perspectives,* Max-Planck-Institute for the History of Science, 149–163.

Hochadel, O. 2013a. 'The multiple Eudald Carbonell: the various roles of Catalonia's most popular archaeologist'. *Dynamis* 33, 2: 389–416.

Hochadel, O. 2013b, *El mito de Atapuerca. Orígenes, ciencia, divulgación* Bellaterra: Edicions UAB.

Hochadel, O. 2013c. 'A boom of bones and books: The "popularization industry" of Atapuerca and human-origins research in contemporary Spain'. *Public Understanding of Science* 22, 5: 530–537.

Hochadel, O. 2013d. 'Palaeoanthropology in the periphery. An introduction'. *Dynamis* 33, 2: 281–295.

Hochadel, O. 2018. 'Wir beseitigen das alte Paradigma! Wie Paläoanthropologen Thomas Kuhn verstehen' in: Feichtinger, J.; Klemun, M.; Svatek, P.; Surman,J. 2018. *Wissenschaftliche Wandlungen und politische Brüche. Festschrift für Mitchell Ash,* Wien.

Hochadel, O.; Carandell, M.; Florensa, C. 2016. 'Scuffles, Scoops and Scams. The Construction of Prehistoric Knowledge in Newspapers'. *Centaurus* 58, 3: 135–256.

Hopke, J. E.; Simis, M. 2015. 'Discourse over a contested technology on Twitter: A case study of hydraulic fracturing'. *Public Understanding of Science*. 26, 1: 105–120.

Howley, A. 2013a. 'Multiple Ancient Hominids Found on Day 2 of Rising Star Expedition' November 12, in: *Voices: Rising Star Expedition* http://voices.nationalgeographic.com/2013/11/16/video-scientists-recover-ancient-hominid-cranium-from-cave/ [Accessed January 2020].

Howley, A. 2013b. 'Video: Scientists Recover Ancient Hominid Skull From Cave' November 16, in: *Voices: Rising Star Expedition* http://voices.nationalgeographic.com/2013/11/16/video-scientists-recover-ancient-hominid-cranium-from-cave/ [Accessed January 2020].

Howley, A. 2013c. 'Video: The Cavers behind South Africa's new hominid discovery' November 15, in: *Voices: Rising Star Expedition* http://voices.nationalgeographic.com/2013/11/15/video-the-cavers-behind-south-africas-new-hominid-discovery/ [Accessed January 2020].

Iglesias, A.; Gibert, J. 2003. 'Orce (1976-2002): Balance de veinticinco años de investigaciones sobre el poblamiento más antiguo de la península ibérica'. *Gallaecia* 22: 21–45.

JagBa, 'Josep Vicenç Santafé i Llopis', in: *Gran Enciclopèdia Catalana*, http://www.enciclopedia.cat/EC-GEC-0245167.xml [Accessed January 2020].

Jiménez-Arenas, J. M.; Santonja, M.; Botella, M.; Palmqvist, P. 2011. 'The oldest handaxes in Europe: fact or artefact?'. *Journal of Archaeological Science* 38, 12: 3340–3349.

Johanson, D.; Maitland, E. 1982. *El Primer antepasado del Hombre* Barcelona: Planeta.

Jones, M. 2013. 'Tooth found in Cave of Orce named earliest human remain' 9 March, in: *The Olive Press* http://www.theolivepress.es/spain-news/2013/03/09/tooth-found-in-cave-of-orce-named-earliest-human-remain/?utm_source=twitter&utm_medium=social&utm_campaign=Tooth+found+in+Cave+of+Orce+named+earliest+human+remain [Accessed January 2020].

Kjærgaard, P. 2011a. 'Ida and Ardi: The fossil cover girls of 2009'. *The Evolutionary Review* 2: 1–9.

Kjærgaard, P. 2011b. 'Hurrah for the missing link!: A history of apes, ancestors and a crucial piece of evidence'. *Notes and Records of the Royal Society* 65, 1: 83–98.

Kjærgaard, P. 2012. 'The fossil trade: Paying a price for human origins'. *Isis* 103, 2: 340–355.

Klein, R. G. 2013. *F. Clark Howell 1925-2007* National Academy of Sciences.

Knorr, K. D.; Krohn, R.; Whitley, R. (eds.) 1981. *The Social Process of Scientific Investigation* Dordrecht: D. Reidel.

Kouper, I. 2010. 'Science blogs and public engagement with science: Practices, challenges, and opportunities'. *Journal of Science Communication* 9(1).

Kuljian, C. 2016. *Darwin's hunch: Science, race and the search for human origins* Johannesburg: Jacana Media.

LaBaracA. 2013. 'Journal of Human Evolution retira temporalmente el artículo sobre el fósil de Orce datado en 1,4 Ma' 17 March, in: *Prehistoria Universal y Vida* http://prehistoria.forosactivos.net/t882-journal-of-human-evolution-retira-temporalmente-el-articulo-sobre-el-fosil-de-orce-datado-en-14-ma [Accessed January 2020].

Lanzarote Guiral, J. M. 2013. 'Dangerous intruder or beneficial influence? The role of the Institut de Paléontologie Humaine in the development of prehistoric archaeology in Spain'. *Complutum* 24, 2: 33–42.

La Otra Mirada 2013, 'El niño de Orce, el enigma' 23 March, in: *Invoox, audiokiosco*, http://www.ivoox.com/otra-mirada-6-temp-el-nino-de-audios-mp3_rf_1892750_1.html [Audio Recording, Accessed January 2020].

Latour, B.; Woolgar, St. 1979. *Laboratory Life: The Social Construction of Scientific Facts* London: Sage.

Latour, B. 1987. *Science in Action: How to Follow Scientists and Engineers through Society* Milton Keynes: Open University Press.

Leakey, R.; Lewin, R. 1977 *Origins. What new discoveries reveal about the emergence of our species and itspossible future*. New York: E.P. Dutton.

Leakey, R.; Lewin, R. 1978. *People of the Lake: Mankind and its Beginnings* New York: Anchor Press/Doubleday.

Leakey, R. 1985. *Leakey* Barcelona: Biblioteca Salvat de Grandes Biografías.

Leibold, J. 2006. 'Competing Narratives of Racial Unity in Republican China: From the Yellow Emperor to Peking Man'. *Modern China* 32, 2: 181–220.

Leitao, L.R. 1978. 'O Passado nas Pedras' *Veja*, 4 October.

Lemonick, M. D. 1993. 'Coming to America'. *Time* May 3: 60–62.

Lewenstein, B. 1992. 'Cold fusion and hot history'. *Osiris* 7: 135–163.

Lewenstein, B. 1995 'From fax to facts: Communication in the Cold Fusion saga'. *Social Studies of Science* 25, 3: 403–436.

Lewin, R. 1997. *Bones of Contention. Controversies in the search for human origins* Chicago: University of Chicago Press.

Liem, T. 2002. *La Osteopatia Craneosacra* Barcelona: Paidotribo.

Lightman, B. 2011. 'Periodicals and Controversy'. *Spontaneous Generations: A Journal for the History and Philosophy of Science* 5, 1: 5–11.

Livingstone, D. N. 2003. *Putting Science in its Place: Geographies of Scientific Knowledge* Chicago: University of Chicago Press.

López Reyes, V.; Martínez Fernández, J. 1997. 'Los conjuntos líticos sobre canto del Paleolítico Medio: El Aculadero' in: Rodríguez Vidal, J. (ed.) *Cuaternario Ibérico* Huelva: Aequa, 326–333.

Lowenstein, J.M. 1981a. 'Immunological reactions from fossil material'. *Philosophical Transactions of the Royal Society of London* B292: 143–149.

Lowenstein, J.M.; Sarich, V.M.; Richardson, B.J. 1981b. 'Albumin systematics of the extinct mammoth and Tasmanian wolf'. *Nature* 291: 409–411.

Lowenstein, J.M.; Borja, C.; García-Olivares, E. 1999. 'Species-specific albumin in fossil bones from Orce, Granada, Spain'. *Human Evolution* 14: 21–28.

Lowenstein, J.M. 2013. 'Jerold M. Lowenstein', in: *Jerold M. Lowenstein, Author, Researcher, Nuclear Physician*. http://www.jeroldlowenstein.com/ [No longer available].

Lozano, C. 'Elecciones Municipales en Andalucía 1979-2015' in: *Història Electoral.com* http://www.historiaelectoral.com/muniand.html [Accessed January 2020].

Lumley, H. de; Lumley, M. de. 1971. 'Discovery of preneanderthal human remains dating from Riss to caune of Arago (Tautavel, Pyrennes orientales)'. *Comptes Rendues des seances de l'Academie de Science* 272, 13: 1729–1742.

Lumley, H. de. 1976. 'Les premières industries humines en Provence' in: Lumley, H. De (ed.) *La Préhistoire Francaise*, Paris: C.N.R.S., 1: 765–76.

Lumley, H. de; Lumley, M. de; Bada, J.; Turekian, K. 1977. 'The Dating of pre-Neandertal remains at Caune de l'Arago, Tautavel, Pyrénées-Orientales. France'. *Journal of Human Evolution* 6, 3: 223–224.

Lumley, H. de (ed.) 1982a. *Les premiers habitants de l'Europe. 1.500.000-100.000 ans* Paris: Laboratoire de Préhistorie du Musée de l'Homme.

Lumley, H. de (ed.) 1982b. *Origine et Évolution de l'Homme* Nice: Laboratoire de Préhistorie du Musée de l'Homme et Action Culturelle Municipale de la ville de Nice.

Lumley, H. de. 1983. 'Els primers habitants d'Europa'. *Revista Avenç* 57: 147–150, (63–66).

Lumley, H. de. 1998. *L'Homme premier* Paris: Odile Jacob.

Lumley, H. de. 2007. *La grande histoire des premiers hommes européens* Paris: Odile Jacob.

Lumley, H. de; Lumley, M. A. de. 2011. *Les premiers peuplements de la Côte d'Azur et de la Ligurie* Colomars: Éditions Melis.

Lynch, M. 1994. *Scientific Practice and Ordinary Action* Cambridge: Cambridge University Press.

Lynch, T. 1990. 'Glacial-age man in South America? a critical review'. *American Antiquity* 55, 1: 12–36.

Lynch, T. 1991. 'Lack of evidence for glacial-age settlement of South America: Reply to Dillehay and Collins and to Gruhn and Bryan'. *American Antiquity* 56, 2: 348–355.

Lyne, J.; Howe, H. F. 1986. ' "Punctuated equilibria": Rhetorical dynamics of a scientific controversy'. *Quaterly Journal of Speech* 72: 132–147.

Machammer, P.; Pera, M.; Baltas, A. 2000. *Scientific controversies. Philosophical and Historical Perspectives*. Oxford: Oxford University Press.

Madariaga de la Campa, B. 2000. *Sanz de Sautuola y el Descubrimiento de Altammira.* Santander: Fundación Marcelino Botin.

Madison, P. 2019. 'All things bleak and bare beneath a brazen sky: practice and place in the analysis of Australopithecus'. *History and Philosphy of Life Sciences* 41(2).

Maerker, A. 2006. 'The tale of the hermaphrodite monkey: classification, state interests and natural historical expertise between museum and court, 1791-1794'. *The British Journal for the History of Science* 39: 29–47.

Maju. 2013. 'An infant's milk tooth from Orce is the earliest known human remain in West Europe' March 2, in: *For what they were ... we are*, http://forwhattheywere-weare.blogspot.com.es/2013/03/an-infants-milk-tooth-from-orce-is.html [Accessed January 2020].

Manias, C. 2014. 'Sinanthropus in Britain: human origins and international science, 1920-1939'. *The British Journal for the History of Science* 48, 2: 289–319.

Martin, B. 1991. *Scientific knowledge in controversy: The social dynamics of the fluoridation debate* Albany: State University of New York Press.

Martin, B. 2014. *Controversy manual* Sparsnäs: Irene Publishing.

Martínez-Navarro, B. 1991. *Revisión sistemática y estudio cuantitativo de la fauna de macromamíferos del yacimiento de Venta Micena (Orce, Granada)* PhD Thesis. Bellaterra: Universitat Autònoma de Barcelona.

Martínez-Navarro, B. 1993. *El Hombre de Orce. Historia de un descubrimiento* Granada: Proyecto Sur.

Martínez-Navarro, B. 1996. 'Similarities between skull fragment VM-0 from Orce (Spain) and the *Homo erectus* holotype from Trinil (Java)'. *Revista Espanola de Paleontologia* 11: 120–121.

Martínez-Navarro, B. 2002. 'The skull of Orce: parietal bones or frontal bones?' *Journal of Human Evolution* 43: 265–270.

Martínez-Navarro, B. 2012. 'The Cranial Fragment from Orce (Granada): A Female Ruminant' *Menga. Revista de Prehistoria de Andalucía* 3: 15–25.

Martínez-Navarro, B. n.d. 'Curriculum Vitae', in: ICREA, *Institució Catalana de Recerca i Estudis Avançats*, https://www.icrea.cat/security/files/researchers/files-maintenance/cvn_bienvenido_martinez_navarro.pdf [Accessed January 2020].

Mascarell, F. 1988. 'La Diputació durant l'etapa Tarradellas: 1977–1980' in: Riquer, Borja de (ed.) *Història de la Diputació de Barcelona* Barcelona: Diputació de Barcelona, 3: 104–129.

McConnell, C. 2000. *The Big Bang-steady state controversy: Cosmology in public and scientific forums* PhD Thesis. Madison: University of Wisconsin.

McMullin, E. 1987. 'Scientific controversy and its termination' in: Engelhardt, H. T.; Caplan, A. L. (eds.) *Scientific Controversies* New York: Cambridge University Press, 49–92.

Meltzer, D. 1989. 'Why don't we know when the first people came to North America?' *American Antiquity* 54, 3: 471–490.

Meltzer, D.; Adovasio, J.; Dillehay, T. 1994. 'On a Pleistocene human occupation at Pedra Furada, Brazil'. *Antiquity* 68: 695–714.

Molina, E. 1998. 'El polémico fósil de Orce: ¿falta de rigor o fraude?' *El Escéptico* 1: 73–76.

Molinero, C.; Ysàs, P. 1988. 'La Diputació de 1949 a 1977' in: Riquer, Borja de (ed.) *Història de la Diputació de Barcelona* Barcelona: Diputació de Barcelona, 3: 42–103.

Molinero, C.; Ysàs, P. 2008. *La anatomía del franquismo. De la supervivencia a la agonía, 1945-1977*. Barcelona: Crítica.

Molinero, C.; Ysàs, P. 2018. *La Transición. Historia y relatos*. Barcelona: Siglo XXI.

Monzón, R. 1989. 'Arqueologia ficción' *Revista de Arqueologia*, X, 98: 6.

Morell, V. 1995. *Ancestral Passions. The Leakey Family quest for humankind's beginnings* New York: Simon and Schuster Inc.

Moreno Fernández, M. L. 2010. *Los primeros catalanes. Jordi, Pau y Lluc, el influjo nacionalista en la búsqueda de los orígenes humanos* Master Thesis. Bellaterra: Universitat Autònoma de Barcelona.

Moreno Lara, V. 2014. *Atapuerca: Arqueología y evolución humana en la prensa* PhD Thesis. Madrid: Universidad Complutense de Madrid.

Moro Abadía, O. 2007 *Arqueología prehistòrica e història de la ciencia* Barcelona: Edicions Bellaterra.

Moro Abadia, O.; Pelayo, F. 2010. 'Reflections on the concept of "precursor": Juan de Vilanova and the discovery of Altamira'. *History of the Human Sciences* 23,4: 1–20.

Morwood, M.; van Oosterzee, P. 2007. *A New Human* USA: Smithsonian Books.

Moyà-Solà, S.; Köhler, M. 1996. 'A Dryopithecus skeleton and the origins of great-ape locomotion', *Nature* 379: 156–159.

Moyà-Solà, S.; Köhler, M. 1997. 'The Orce skull: anatomy of a mistake' *Journal of Human Evolution* 33: 91–97.

Moyà-Solà, S. et al. 2004. 'Pierolapithecus catalaunicus, a new Middle Miocene great ape from Spain'. *Science* 306: 1339-1344.

Moyà-Solà, S.; Pons, Guillem X. 2007. 'Rafel Adrover, paleontòleg, (1911–2007) in memoriam'. *Bolletí de la Societat d'Història Natural de les Balears* 49: 213–218.

Mulkay, M.; Gilbert, N. 1981. 'Putting Philosophy to Work: Karl Popper's Influence on Scientific Practice' *Philosophy of the Social Sciences*, 11: 389–407.

Muniesa, B. 2005. *Dictadura y transición: La monarquía parlamentaria* Barcelona: Publicacions i Edicions de la Universitat de Barcelona.

Myers, G. 2003. 'Discourse studies of scientific popularisation: questioning the boundaries'. *Discourse Studies* 5, 2: 265–279.

Navarro Espinosa, M. 2010. *El Hombre de Orce* La Nave de Tharsis S.L. [Documentary].

Naylor, S. K. 2005. 'Introduction: Historical Geographies of Science-Places, Contexts, Cartographies'. *British Journal for the History of Science* 38: 1–12.

Nieto-Galan, A. 2011. *Los públicos de la ciencia* Madrid: Ambos Mundos, Marcial Pons.

Niño de Orce, El. 2013a. 'Far away, far away'. 8 March, in: *El Niño de Orce, Historia del Alfa i el Omega*, http://elnenedeorce.blogspot.com.es/2013/03/faraway.html [No longer available].

Niño de Orce, El. 2013b. 'El diente incorrupto de Santo Hominido de Orce'. 16 March, in: *El Niño de Orce, Historia del Alfa i el Omega*, http://elnenedeorce.blogspot.com.es/2013/07/el-diente-incorrupto-de-santo-hominido.html [No longer available].

Niño de Orce, El. 2013c. 'Orce no te adora'. 31 July, in: *El Niño de Orce, Historia del Alfa i el Omega*, http://elnenedeorce.blogspot.com.es/2013/07/orce-no-te-adora.html [No longer available].

Niño de Orce, El. 2013d. 'Poca chicha y mucho hueso'. 5 March, in: *El Niño de Orce, Historia del Alfa i el Omega*, http://elnenedeorce.blogspot.com.es/2013/03/ya-ha-nacido.html [No longer available].

Niño de Orce, El. 2013e. 'Quienes somos, de dónde venimos'. 22 March, in: *El Niño de Orce, Historia del Alfa i el Omega*, http://elnenedeorce.blogspot.com.es/2013/03/quienes-somos-de-donde-venimos.html [No longer available].

Niño de Orce, El. 2013f. 'Sin Regreso a Ítaca'. 6 March, in: *El Niño de Orce, Historia del Alfa i el Omega*, http://elnenedeorce.blogspot.com.es/2013/03/sin-regreso-itaca.html [No longer available].

Niño de Orce, El. 2013g. 'Robos Culturales, dientes para recuerdo'. 19 March, in: *El Niño de Orce, Historia del Alfa i el Omega*, http://elnenedeorce.blogspot.com.es/2013/03/robos-culturales-dientes-para-recuerdo.html [No longer available].

Niño de Orce, El. 2013h. 'Motivos de la Retirada, opinioques'. 20 March, in: *El Niño de Orce, Historia del Alfa i el Omega*, http://elnenedeorce.blogspot.com.es/2013/03/y-hete-aqui-los-motivos-de-la-retirada.html [No longer available].

Niño de Orce, El. 2013i. 'Un disparate – Once años perdidos'. 25 March, in: *El Niño de Orce, Historia del Alfa i el Omega*, http://elnenedeorce.blogspot.com.es/2013/03/un-disparate.html [No longer available].

Niño de Orce, El. 2013j. 'Troles son amores, y no buenas razones'. 21 March, in: *El Niño de Orce, Historia del Alfa i el Omega*, http://elnenedeorce.blogspot.com.es/2013/03/troll.html [No longer available].

Niño de Orce, El. 2013k. 'Las Flores de Mayo – Regresa El Niño del Journal'. 10 May in: *El Niño de Orce, Historia del Alfa i el Omega*, http://elnenedeorce.blogspot.com.es/2013/05/las-flores-de-mayo-regresa-el-nino.html [No longer available].

Niño de Orce, El. 2013l. '¿Cueva or "Cave"? No lo "sabe" '. 9 March, in: *El Niño de Orce, Historia del Alfa i el Omega*, http://elnenedeorce.blogspot.com.es/2013/03/cueva-or-cave-no-lo-sabe.html [No longer available].

Niño de Orce, El. 2013m.'Temporalmente retirado – ¿una caries?'. 19 March, in: *El Niño de Orce, Historia del Alfa i el Omega*, http://elnenedeorce.blogspot.com.es/2013/03/temporalmente-eliminado.html [No longer available].

Nomade, S. et al. 2011. 'First 40Ar/39Ar age of the Ceprano man (central Italy)' *Quaternary Geochronology* 6, 5: 453–457.

Olesko, K. M. 2007. "The world we have lost: history as art" *Isis*, vol. 98. No. 4, pp. 760–768.

Oransky, I. 2013. 'Toothless wonder? Paper on "oldest human fossil in Europe" temporarily removed from journal's site' 19 March, in: *Retraction Watch*, http://retractionwatch.com/2013/03/19/toothless-wonder-paper-on-oldest-human-fossil-in-europe-temporarily-removed-from-journals-site/ [Accessed January 2020].

Palmqvist, P.; Gibert, J.; Martfnez, B. 1992. 'Sobre la relación tamaño/abundancia en los macromamíferos de Venta Micena y sus implicaciones tafonómicas'. *Revista Española de Paleontología* 7: 174–180.

Palmqvist, P. 1997. 'A critical re-evaluation of the evidence for the presence of hominids in lower Pleistocene times at Venta Micena, Southern Spain'. *Journal of Human Evolution* 33: 83–89.

Papanelopoulou, F.; Nieto-Galan, A.; Perdiguero E. (eds.) 2009. *Popularizing Science and Technology in the European Periphery, 1800-2000* London: Ashgate.

Pedregosa, F. 1983. 'Orce, dos millones de años después' *Revista Tiempo*, 69: 50–51.

Phillips, P. 1980. *The prehistory of Europe* London: Allen Lane.

Pinch, T. 1994. 'Cold fusion and the sociology of scientific knowledge'. *Technical Communication Quarterly* 3, 1: 85–100.

Piqueras, M. 2013a. 'L'article sobre la dent de llet d'Orce, retirat temporalment'. 19 March, in: *La Lectora Corrent*, http://lectoracorrent.blogspot.com/2013/03/larticle-sobre-la-dent-de-llet-dorce.html [Accessed January 2020].

Piqueras, M. 2013b. 'Continua la polèmica sobre l'article que descriu la dent trobada a Orce'. 23 March, in: *La Lectora Corrent*, http://lectoracorrent.blogspot.com/2013/03/continua-la-polemica-sobre-larticle-que.html [Accessed January 2020].

Piqueras, M. 2013c. 'Orce, Gibert i 1984'. 20 May, in: *La Lectora Corrent*, http://lectoracorrent.blogspot.com/2013/05/orce-gibert-i-1984.html [Accessed January 2020].

Pitts, M.; Roberts, M. 1997. *Fairweather Eden* London: Century.

Podgorny, I. 2005. 'Bones and devices in the Constitution of Paleontology in Argentina at the End of the Nineteenth Century'. *Science in Context* 18, 2: 249–283.

Pons Moyà, J. 1985. 'Nota preliminar sobre el hallazgo de *Homo* sp. En los rellenos cársticos de Cueva Victoria'. *Endins*, 10–11: 47–50.

Prager, E.M.; Wilson, A.C.; Lowenstein, J.M.; Sarich, V.M. 1980. 'Mammoth albumin'. *Science* 209: 287–289.

Prous, A.; Fogaça, E. 1999. 'Archaeology of the Pleistocene-Holocene boundary in Brazil'. *Quaternary International*, 53–54: 21–41.

Pruitt, T. C. 2011. *Authority and the Production of Knowledge in Archaeology* PhD Thesis. Cambridge: University of Cambridge.

Pyne, L.V. 2016. *Seven skeletons: The evolution of the world's most famous human fossils* New York: Viking.

Querol, M.A.; Santonja, M. 1983. 'El Yacimiento de El Aculadero (Puerto de Santa María, Cádiz)'. *Revista de Arqueología* 29: 8–16.

Redacción Algo. 1984. 'Orce, un año después'. *Revista Algo* June: 60–61.

Redacción El Jueves. 1984. 'Gran júbilo entre los monos'. *El Jueves* 365.

Redacción El Papus. 1984. 'El Misterio del Hombre de Orce'. *El Papus* 525: 1.

Redacción Veja. 1983. 'Idade do Carvão'. *Veja* 11 May: 64.

Redacción Veja. 1986. 'Teoria Revista'. *Veja* 13 August: 87.

Redacción Veja. 1987. 'O americano perdido'. *Veja* 30 September: 61–62.

Reumer, J.W. F.; De Vos, J. 1999. 'Short biography and a bibliography of Paul Y. Sondaar' in: Reumer, J.W. F.; De Vos, J. (eds.) *Elephants have a snorkel! Paper in honour of Paul Y. Sondaar,* Deinsea 7: 3–19, Rotterdam: Natuurmuseum Rotterdam.

Richmond, J. 2009. *Experts and Australopithecines: Credibility and controversy in the science of human evolution, 1924–1959* PhD Thesis. San Diego: University of California San Diego.

Riquer, B. de. (ed.) 1988. *Història de la Diputació de Barcelona* Barcelona, Diputació de Barcelona.

Rius, J.; Martínez, S. 2012. 'La política cultural de la Diputación de Barcelona. La influencia del contexto político institucional y de la articulación entre administraciones en el desarrollo de un modelo singular de cooperación cultural'. *RIPS: Revista de Investigaciones Políticas y Sociológicas* 11, 1: 9–37.

Roberts, M.B.; Stringer, C.B.; Parfitt, S.A. 1994. 'A hominid tibia from Middle Pleistocene sediments at Boxgrove, UK'. *Nature* 369: 311–313.

Rödder, S.; Franzen, M.a; Weingart, P., (eds.) 2012. T*he sciences' media connection – communication to the public and its repercussions* Dordrecht: Springer.

Roebroeks, W.; Van Kolfschoten, T. 1994. 'The earliest occupation of Europe: a short chronology'. *Antiquity* 68: 489–503.

Roebroeks, W.; Van Kolfschoten, T. (eds.) 1995a. *The earliest occupation of Europe. Proceedings of the European Science Foundation workshop at Tautavel (France), 1993*. Leiden: University Press.

Roebroeks, W.; Van Kolfschoten, T. 1995b. 'The earliest occupation of Europe: a reappraisal of artefactual and chronological evidence' in: Roebroeks; Van Kolfschoten (eds.) *The earliest occupation of Europe. Proceedings of the European Science Foundation workshop at Tautavel (France)*, 1993. Leiden: Leiden University Press.

Romero, A. 1984. 'Chernenko desciende del Burro'. *El Jueves* 365: 3.

Rosa de los Vientos 2013, 'El hallazgo del Hombre de Orce' 4 March, Onda Cero. [Audio Recording].

Rudwick, M. 1985. *The Great Devonian Controversy: The Shaping of Scientific Knowledge among Gentlemanly Specialists* Chicago and London: The University of Chicago Press.

Ruiz Zapatero, G. 1996. 'La divulgación del pasado. Arqueólogos y periodistas: una relación posible'. *Boletín del Instituto Andaluz de Patrimonio Histórico* 17: 96–99.

Ruiz Zapatero, G.; Castaño, A. M. 1999. 'L'arqueologia en els mitjans de comunicació: materials per a una reflexió crítica sobre la divulgació del passat'. *Cota Zero: Revista d'Arqueologia* 15: 42–62.

Sánchez, F.; Gibert, J.; Malgosa, A.; Ribot, F.; Gibert, L.; Walker, M. 1999. 'Insights into the evolution of child growth from Lower Pleistocene humeri at Venta Micena (Orce, Granada province, Spain)'. *Human Evolution* 14, 1–2: 63–82.

Sautman, B. 2001. 'Peking man and the politics of paleoanthropological nationalism in China'. *Journal of Asian Studies* 60, 1: 95–124.

Schlanger, N. 2002. 'Ancestral Archives: Explorations in the History of Archaeology'. *Antiquity* 76: 127–131.

Schmalzer, S. 2008. *The People's Peking Man: Popular Science and Human Identity in Twentieth-Century China*. Chicago and London: The University of Chicago Press.

Schweitzer, M. et al. 2009. 'Biomolecular characterization and protein sequences of the Campanian hadrosaur B. canadensis'. *Science* 324: 626–631.

Scott, G. R., & Gibert, L. (2009). 'The oldest hand-axes in Europe'. *Nature*, 461(7260), 82–5.

Secord, J. A. 2002 *Victorian Sensation: The Extraordinary Publication, Reception and Secret Authorship of 'Vestiges of the Natural History of Creation'* Chicago: University of Chicago Press.

Secord, J. A. 2004. 'Knowledge in transit'. *Isis* 95: 654–672.

Shapin, S. 1995. 'Cordelia's Love: Credibility and the Social Studies of Science'. *Perspectives on Science* 3: 255–275.

Shapiro, H. L. 1974. *Peking Man*. New York: Simon and Schuster.

Shinn, T.; Cloître, M. 1985. 'Expository practice: social, cognitive and epistemological linkage' in: Shinn, T.; Whitley, R. (eds.) *Expository science: forms and functions of popularisation* Dordrecht: Reidel.

Shinn, T.; Whitley, R. (eds.) 1985. *Expository science: forms and functions of popularisation* Dordrecht: Reidel.

Siniestro Total, '¿Quiénes somos? ¿De dónde venimos? ¿A dónde vamos?'. Dro, 2002. [Music Record].

Simandiraki-Grimshaw, S. 2012. *From Archaeology to Archaeologies: The "Other" Past* Archaeopress: Oxford.

Sloan, C. 2004. *The Human Story* Washington: National Geographic.

Sommer, M. 2006. 'Mirror, Mirror on the Wall: Neanderthal as Image and "Distortion" in Early 20th-Century French Science and Press'. *Social Studies of Science* 36, 2: 207–240.

Sommer, M. 2007. *Bones and Ochre: The Curious Afterlife of the Red Lady of Paviland* Cambridge, Mass./London: Harvard University Press.

Sommer, M. 2008. 'History in the Gene: Negotiations between Molecular and Organismal Anthropology'. *Journal of the History of Biology* 41, 3: 473–528.

Soto Fernández, D.; Infante-Amate, J.; Villa, I.; Jaén, S. 2015. *La identidad andaluza como identidad cívica durante la construcción de la autonomía (1975-1982)* Sevilla: Centro de Estudios Andaluces, Junta de Andalucía.

Spencer, F. 1990. *Piltdown: A Scientific Forgery* New York: Oxford University Press.

Staley, R. 1998. 'On the histories of relativity: the propagation and elaboration of relativity theory in participant histories in Germany, 1905-1911'. *Isis* 89, 2: 263–299.

Theolduvaigorge. 2013. 'The oldest human fossil in Europe dated to ca. 1.4 Ma at Orce (Spain)' 7 March, in: *None of Your Neurones Know Who You Are* ... http://theolduvaigorge.tumblr.com/post/44805249011/the-oldest-human-fossil-in-europe-dated-to-ca-1-4 [Accessed January 2020].

Thibault, C.; Querol, M.A.; Viguier, C.; Santoja, M. 1977. 'El yacimiento del Paleolítico inferior arcaico de El Aculadero" (Puerto de Santa María, Cadiz)' in: Congresos Arqueológicos Nacionales, Secretaría General (ed.) *XIV Congreso Nacional de Arqueología (Vitoria, 1975)*, 69–80.

Tixier, J.; Roe, D.; Turq, A.; Gibert, J.; Martínez-Navarro, B.; Arribas, A.; Gibert, L.; Maillo, L.; Iglesias, A. 1995. 'Présence d'industrie lithique dans le Pleistocène de la région d'Orce (Grenada, Espagne): quel est l'état de la question?'. *Comptes Rendus de l'Academie des Sciences de Paris* 321: 71–78.

Tobias, P. V. 1998. 'Some comments on the case for Early Pleistocene hominids in South-Eastern Spain'. *Human Evolution*, 13, 2.: 91–96.

Tobias, P.V. 2005. *Into the past: a memoir* Johannesburg: Picador Africa.

Toro-Moyano, I.; Martínez-Navarro, B.; Agustí, J.; Souda, C.; Bermúdez de Castro, J. M.; Martinón-Torres, M.; Fajardo, B.; Duval, M.; Oms, O.; Parés, J. M.; Julià, R.; Moigne, A.M.; Espigares, M. P.; Ros-Montoya, S.; Palmqvist, P. 2013a. 'The oldest human fossil in Europe dated to ca. 1.4 Ma at Orce (Spain)' *Journal of Human Evolution* 1–9. In press, original version.

Toro-Moyano, I.; Martínez-Navarro, B.; Agustí, J.; Souda, C.; Bermúdez de Castro, J. M.; Martinón-Torres, M.; Fajardo, B.; Duval, M.; Oms, O.; Parés, J. M.; Julià, R.; Moigne, A.M.; Espigares, M. P.; Ros-Montoya, S.; Palmqvist, P. 2013b. 'The oldest human fossil in Europe, from Orce (Spain)'. *Journal of Human Evolution* 65, 1: 1–9.

Torreiglesias, M. 1986. 'Los Fósiles son patrimonio de la humanidad'. *Muy Interesante* 60: 129–131.

Trench, B. 2012. 'Scientists' Blogs: Glimpses Behind the Scenes' in: Rödder, S.; Franzen, M.; Weingart, P., (eds.) *The sciences' media connection – communication to the public and its repercussions* Dordrecht: Springer.

Trigger, B. G. 1992. *Historia del pensamiento arqueológico* Barcelona: Crítica.

Turq, A.; Martinez-Navarro, B.; Palmquist, P.; Arribas, A.; Agustí, J.; Rodríguez Vidal, J. 1996. 'Le Plio-Pléistocène de la région d'Orce, Province de Grenade, Espagne: Bilan et perspectives de recherche'. *Paleo* 8: 161–204.

Turner, C. G. 1986. 'Science and the citizen'. *Scientific American* 255 3: 68–69.

TVE 2011, 'Paso del Estrecho de Gibraltar'. *Arqueomanía*, Episode 16, 4 December. [Video Recording].

TVE 1983, La Tarde. [Video Recording].

TV1 1995a, Telesur, 1 September. [Video Recording].

TV1 1995b, Telesur, 10 September. [Video Recording].

Ucko, P. 1987. *Academic Freedom and Apartheid: The story of the World Archaeological Congress* Londres: Duckworth.

UHF 1983, La Clave. [Video Recording].

Valoch, K. 1976 'Aperçu des premières industries en Europe' in: Valoch, K. (ed.) *IX Congrès, Colloque VIII, Les premières industries de l'Europe* Nice: Union Internationale des sciences Préhistoriques et Protohistoriques, 178.

Van Dijck, M. 2008. 'From science to popularisation and back – the science and journalism of the Belgian economist Gustave de Molinari'. *Science in Context* 21, 3: 377–402.

Van Oosterzee, P. 1999. *The Story of Peking Man* Crows Nest, Australia: Allen and Unwin.

Van Reybrouck, D. 2002. 'Boule's error: on the social context of scientific knowledge'. *Antiquity* 76: 158–164.

Van Reybrouck, D.; de Bont, R.; & Rock, J. 2009. 'Material Rhetoric: Spreading Stones and Showing Bones in the Study of Prehistory'. *Science in Context* 22, 2: 195–216.

Veltri, G. 2012. 'Microblogging and nanotweets: Nanotechnology on Twitter' *Public Understanding of Science* 22, 7: 832–849.

Wagenknecht, S. 2012. 'Debating Troy in the Mass Media – The Catalytic Impact of Public Controversy on Academic Discourse' in: Rödder, S.; Franzen, M.; Weingart, P. (eds.) 2012. *The sciences' media connection – communication to the public and its repercussions* Dordrecht: Springer, 291–306.

Walker, M.; Gibert, J. 1992. 'Cabezo Gordo (Torre Pacheco-Murcia): investigaciones preliminares de 1992' *Memorias de Arqueologia* 7.

Walker, M.J.; Gibert, J.; López, M.; Lombardi. V.; Pérez Pérez, A.; Zapata, J.; Ortega, J.; Higham, T.; Pike, A.; Schwenninger, J-L.; Zilhao, J.; Trinkaus, E. 2008 'Late Neandertals in Southeastern Iberia: Sima de las Palomas del Cabezo Gordo, Murcia, Spain'. *Proccedings of the National Academy of Sciences* 105: 20631–20636.

Walsh, J. E. 1996. *Unraveling Piltdown: The Science Fraud of the Century and Its Solution* New York: Random House.

Warren, M. 2019. 'Biggest Denisovan fosil yen sppills ancient human's secrets' *Nature*, 1 May 2019.

Weingart, P. 1998. 'Science and the media'. *Research Policy* 27: 869–879.

Wellbourne, D. J.; Grant, W. J. 2015. 'Science communication on YouTube: Factors that affect channel and video popularity' *Public Understanding of Science* 19 February.

Whitley, D.; Dorn, R. 1993. 'New perspectives on the clovis vs. Pre-clovis controversy'. *American Antiquity* 58, 4: 626–647.

Wikipedia Contributors, 'Partit Socialista Unificat de Catalunya', in: *Wikipedia, The Free Encyclopedia*, http://ca.wikipedia.org/wiki/Partit_Socialista_Unificat_de_Catalunya [Accessed January 2020].

Wikipedia Contributors, 'Partit dels i les Comunistes de Catalunya', in: *Wikipedia, The Free Encyclopedia*, http://ca.wikipedia.org/wiki/Partit_dels_i_les_Comunistes_de_Catalunya [Accessed January 2020].

Wikipedia Contributors, 'Estatut d'Autonomia de Catalunya de 1979', in: *Wikipedia, The Free Encyclopedia*, http://ca.wikipedia.org/wiki/Estatut_d%27Autonomia_de_Catalunya_de_1979 [Accessed January 2020].

Wikipedia Contributors, 'Estatuto de Autonomía de Andalucía 1981', in: *Wikipedia, The Free Encyclopedia*, http://es.wikisource.org/wiki/Estatuto_de_Autonom%C3%ADa_de_Andaluc%C3%ADa_1981 [Accessed January 2020].

Wikipedia Contributors, 'Orce', in: *Wikipedia, The Free Encyclopedia*, http://es.wikipedia.org/wiki/Orce [Accessed January 2020].

Wikipedia Contributors, 'Pierre Mein', in: *Wikipedia, The Free Encyclopedia*, https://ca.wikipedia.org/wiki/Pierre_Mein [Accessed January 2020].

Wikipedia Contributors, 'Never in the field of human conflict was so much owed by so many to so few', in: *Wikipedia, The Free Encyclopedia*, http://en.wikipedia.org/wiki/Never_was_so_much_owed_by_so_many_to_so_few [Accessed January 2020].

Wikipedia Contributors, 'Mark Zuckerberg', in: *Wikipedia, The Free Encyclopedia*, http://es.wikipedia.org/wiki/Mark_Zuckerberg#cite_note-10 [Accessed January 2020].

Wikipedia Contributors, 'Rising Star Cave', in: *Wikipedia, The Free Encyclopedia*, http://en.wikipedia.org/wiki/Rising_Star_Cave [Accessed January 2020].

Wikipedia Contributors, 'Spanish Transition to Democracy' in: *Wikipedia, The Free Encyclopedia*, https://en.wikipedia.org/wiki/Spanish_transition_to_democracy#The_PSOE_in_government [Accessed January 2020].

Wikipedia Contributors, 'Henry de Lumley', in: *Wikipedia, The Free Encyclopedia*, http://fr.wikipedia.org/wiki/Henry_de_Lumley [Accessed January 2020].

Wikipedia Contributors, 'Musée de paléontologie humaine de Terra-Amata', in: *Wikipedia, The Free Encyclopedia*, https://fr.wikipedia.org/wiki/Mus%C3%A9e_de_pal%C3%A9ontologie_humaine_de_Terra-Amata [Accessed January 2020].

Wikipedia Contributors, 'Alfredo Relaño', in: *Wikipedia, The Free Encyclopedia*, https://es.wikipedia.org/wiki/Alfredo_Rela%C3%B1o#cite_note-2 [Accessed January 2020].

Wikipedia Contributors, 'Manuel Fraga Iribarne', in: *Wikipedia, The Free Encyclopedia*, http://ca.wikipedia.org/wiki/Manuel_Fraga_Iribarne [Accessed January 2020].

Wikipedia Contributors, 'Gerhard Bosinski', in: *Wikipedia, The Free Encyclopedia*, http://de.wikipedia.org/wiki/Gerhard_Bosinski_%28Pr%C3%A4historiker%29 [Accessed January 2020].

Wikipedia Contributors, 'Milford H. Wolpoff', in: *Wikipedia, The Free Encyclopedia*, https://en.wikipedia.org/wiki/Milford_H._Wolpoff [Accessed January 2020].

Willis, D. 1992. *The Leakey Family: Leaders in the Search for Human Origins* New York: Library of Congress.

Woolgar, S. 1976. 'Writing an intellectual history of scientific development: the use of discovery accounts'. *Social Studies of Science* 6, 3: 395–422.

Yeo, S., et al. 2016. 'The case of #arseniclife: Blogs and Twitter in informal peer review' *Public Understanding of Science*, Vol. 26, 8, 937–952.

Zihlman, A.; Lowenstein, J. 1996. 'A Spanish Olduvai?' *Current Anthropology* 37, 4: 695–697.

4 The Press

Ache, J., *Diari de Sabadell* 1983, 4 June. 'Hallados de los restos del Hombre más antiguo de Europa': 1–3.

Ache, J., *Diari Sabadell* 1995, 7 September. 'La polèmica d'Orce segueix a l'Institut de Paleontologia': 8.

A.G., *ABC* 2013, 2 March. 'Descubren en Orce el diente fósil más antiguo de Europa': 63.

Aguilar, J., *El País* 1983, 14 June. 'El hallazgo del hombre de Orce, puede suponer una revolución en el estudio de la especie humana'.

Agustí, J., *La Vanguardia* 1983, 21 August 'Crònica de un hallazgo'.

Agustí, J.; Gibert, J.; Moyà-Solà, S., *La Vanguardia* 1983, 21 August. 'Características del Hombre de Orce'.

Amela, V. *La Vanguardia* 1985, 2 Feburary. 'Inaugurada en las Drassanes de Barcelona la exposición "Origen I Evolució de l'Home" ': 25.

A.M.M., *ABC* 1985, 22 August. 'Profesor Gibert: "Mantengo que los huesos de Orce son de naturaleza humana" ': 40.

Andrés, J., *El País* 1984, 18 April. 'Los restos líticos de Castellón evidencian la presencia de "Homo habilis" en Europa'.

Ansede, M., *Materia* 2013a, 12 March. 'Así vivió el primer europeo conocido'.

Ansede, M., *Materia* 2013b, 20 March. 'Retirado "temporalmente" el estudio sobre el primer europeo conocido'.

Antón, J., *El País* 1993, 13 May. 'El Hombre de Orce acaba su "travesía del desierto", dice su descubridor'.

Antón, J., *El País* 1994, 26 January. 'Lumley descalifica al Hombre de Orce'.

Antón, J., *El País* 2000, 27 January. 'Notícies de la Prehistòria': Quadern 1–3.

Antón, J., *El País* 2005, 31 July. 'Pau, el abuelo en la basura'.

Ap y Efe, *La Vanguardia* 1986, 2 September. 'Southampton: tórmenta política sobre el congreso mundial de arqueología que empezó ayer': 33.

Aragay, I., *Avui* 1995, 2 September. 'Orce, el nou paradigma científic': B1.

Aranda, G., *El Mundo* 2014, 27 July. 'Piauí: tras las huellas del primer hombre americano'.

Arias, J., *El País* 1995a, 2 September. 'El descubridor del "Hombre de Orce" anuncia nuevos hallazgos anteriores a Atapuerca'.

Arias, J., *El País* 1995b, 5 September. 'Expertos creen que los fósiles de Orce son anteriores a los de Atapuerca'.

Arias, J., *El País* 1995c, 11 September. 'Orce y la búsqueda de la cuna de los europeos'.

Arias, J., *El País* 2000a, 1 September. 'Hallados en Orce restos de animales mamíferos de hace 4 millones de años'.

Arias, J., *El País* 2000b, 8 July. 'Investigan restos de mamíferos de hace cuatro millones de años en Baza'.

Arias, J., *El País* 2000c, 21 November. 'Una muestra recorre en Granada la prehistoria del hombre europeo'.

Arias, J., *El País* 2000d, 2 December. 'No tiene por qué haber rivalidad entre Orce y Atapuerca'.

Arias, J., *El País* 2002a, 14 March. 'Cultura deniega a Josep Gibert el permiso para excavar en Orce'.

Arias, J., *El País* 2002b, 19 March. 'La Junta rechazó la solicitud de Gibert de excavar en Orce por motivos científicos y administrativos'.

Arroyo, F., *El País* 1984, 13 May. 'Josep Gibert considera que la hipótesis más fidedigna sigue siendo la del homínido'.

Avila Granados, J. *La Vanguardia* 1984, 21 November. 'Venta Micena es el yacimiento con mejor historia geológica de Europa, según el doctor Josep Gibert': 33.

A.V.G., *El País* 2001, 16 July. 'Los paleontólogos hallan en Orce el asta de un ciervo primitivo que vivió hace 1,2 millones de años'.

Barata, F., *El Periódico* 1983, 29 June.'La Diputación ayudará a los investigadores de Orce':19.

Barata, F., *El Periódico* 1984, 11 January. 'Los paleontologos hallan utensilios del Hombre de Orce'.

Barata, F., *El Periódico* 1985, 2 February. 'El fósil de Múrcia es 100.000 años más viejo que el de Orce': 15.

Barata, F., *El Periódico* 1986, 17 December. 'Científicos de todo el mundo reconocen al Hombre de Orce': 32–33.

Barata, F., *El Periódico* 1987a, 2 November. 'Dos científicos afirman que el cráneo de Orce es de un equino': 24.

Barata, F., *El Periódico* 1987b, 4 October. 'El descubridor del cráneo de Orce insiste en que es humano': 27.

Bedinelli, T., *El País* 2014, 18 May. 'Piauí: un gigante para la investigación científica, invisible para Brasil'.

Belik, H., *Folha de São Paolo* 1983, 1 May. 'Pinturas rupestres do piaui derrumbam teoria': 30.

Belik, H., *Folha de São Paolo* 1984, 23 September. 'No Piauí, o maior sitio arqueólogico das Americas': 26.

Bonalume, R., *Folha de São Paolo* 1986, 25 January. 'Betty Meggers inicia polêmica com os cientistas brasileiros'.

Bonalume, R., *Folha de São Paolo* 2009, 12 December. 'Machadinha de 900 mil anos é amais antiga da Europa'.

Bonet, E., *La Opinión de Granada* 2007, 2 March. 'Corriente Alterna'.

Cambril, A., *La Opinión de Granada* 2007, 21 January. 'El Burro Antecesor': 3.

Campillo, D., *El País* 1987, 22 October. 'El Cráneo de Orce'.

Cappa, G., *Granada Hoy* 2007, 23 June. 'Cultura niega el permiso para la gran excavación prevista en Orce'.

Carbonell, E. *El Periódico* 1995, 11 September. 'Apunte': 42.

Castro, E., *El País* 1983, 28 June. 'Los restos del hombre de Orce llegaron ayer al Instituto de Paleontologia de Sabadell'.

Cervera, J., *El País* 1995, 11 September. 'La difícil ruta desde África'.

Clemyn-Jones, B., *The Guardian* 1983, 15 June. 'Skull sheds new light in mankind': 6.

Colomer, V., *Diari de Sabadell* 1984, 19 May. 'Salvador Moyà, ¿hombre?, de Orce': 5.

Contreras, M., *El País* 1985, 31 January. 'El equipo que descubrió el Hombre de Orce halla en Murcia "el resto humano más antiguo de Europa"'.

Corbella, J.; Sáez, C., *La Vanguardia* 2008, 21 July. 'La prehistoria se revela en verano': 26.

Cortés, V., *El País* 2013, 16 March. 'El diente de "oro" de Orce'.

Dalmau, A., *La Vanguardia* 1984, 5 January. 'Un año de transición y de reforma para la Diputación de Barcelona': 16.

Da Reportagem Local, *Folha do São Paolo* 1986, 25 June. 'Piauí terá un museu para achados arqueológicos': 16.

Da Reportagem Local, *Folha do São Paolo* 1993, 11 April. 'Francês não acredita no Homem do Piauí': 6.

Das Agéncias Internacionals, *Folha do São Paolo* 1986, 19 June. 'Nature divulga descobertas feitas no Piauí'.

Debelius, H. *The Times* 1983, 15 June. 'No title': 3.

Debelius, H. *The Times* 1984, 14 May. 'Scientists make ass of an old skull': 10.

De León-Sotelo, T., *ABC* 1994, 31 July. 'La "resurrección" del "hombre de Orce"': 50.

De León-Sotelo, T., *ABC* 1995a, 14 August. 'Investigadores han hallado cientos de fósiles que avalan que en Orce habitó el primer europeo': 51.

De León-Sotelo, T., *ABC* 1995b, 2 September. 'José Gibert: "El congreso de Orce cambiará la prehistoria en Europa"': 53.

De León-Sotelo, T., *ABC* 1996a, 23 August. 'El alcalde de Orce expulsa al científico que no cree que el fragmento craneal es de un homínido': 41.

De León-Sotelo, T., *ABC* 1996b, 4 August. 'El equipo del Hombre de Orce va de cráneo': 49.

De León-Sotelo, T., *ABC* 1996c, 8 August. 'Orce, una polémica que no fosiliza': 47.

De León-Sotelo, T., *ABC* 1996d, 24 August. 'El alcalde de Orce exige que regrese al pueblo el polémico fragmento craneal': 49.

De León-Sotelo, T., *ABC* 1996e, 3 September. 'Andalucía, dispuesta a ir a los tribunales si Sabadell no devuelve los fósiles de Orce': 69.

Del Hoyo, M. Á., *El Periódico* 1995a, 5 September. 'Los primeros Europeos vivieron en Orce un millón de años antes que en Atapuerca': 20.

Del Hoyo, M Á. *El Periódico* 1995b, 8 September. 'El Hombre de Orce' aceptado como el más viejo de Europa': 21.

Del Hoyo, M. A.; Madridejos, A., *El Periódico* 1995, 11 September. 'Orce y Atapuerca remueven la prehistoria de Europa': 42.

De Renzi, M.; Porta, J.; Truyols, J., *Avui* 1985, 6 March. 'Nota sobre paleontologia': 10.

De Semir, V., *La Vanguardia* 1983, 12 June. 'Tres paleontólogos catalanes descubren el hombre más antiguo de Europa y Asia': 51.

De Torres, C., *La Vanguardia* 1986, 31 August. 'Los primeros pobladores del continente americano pudieron llegar hace más de 30.000 años'.

Editorial department, *Gadsen Times* 1984, 13 May. 'Orce Man may have been animal'.

Editorial department, *Ottawa Citizen* 1984, 14 May. 'Oldest human remains might be from donkey'.

EFE, *ABC* 1983, 30 June. 'El Hombre de Orce volverá a Andalucía': 42.

EFE, *ABC* 1984a, 13 May. 'Josep Gibert niega que el cráneo de Orce sea de un asno': 51.

EFE, *ABC* 1984b, 14 May. 'No habrá respuesta oficial sobre el Hombre de Orce hasta diciembre': 29.

EFE, *ABC* 1990a, 12 February. 'Hallados nuevos restos del "Hombre de Orce" que eliminan dudas sobre su carácter humano': 42.

EFE, *ABC* 1990b, 19 August. 'El paleontólogo José Gibert asegura haber encontrado nuevos restos del Hombre de Orce': 43.

EFE, *ABC* 1995, 9 September. 'La Junta de Andalucía financiará y protegerá el yacimiento de Orce': 56.

EFE, *ABC* 1997, 18 March. 'Gibert pide garantías para el traslado del hombre de Orce': 56.

EFE, *ABC* 2005, 30 July. 'Hallan en Orce restos de un mamut con una antigüedad de 1,2 millones de años': 64.

EFE, *El Ideal de Granada* 2013, 27 January. 'Se reactivan los estudios en los yacimientos de Orce (Granada)'.

EFE, *El Mundo* 2007, 12 October. 'Josep Gibert Clols, descubridor del primer poblador de Europa Occidental'.

EFE, *El País* 2001, 23 July. 'Hallada en Orce una pelvis de elefante de 1,2 millones de años'.

EFE, *El País* 2007, 16 January. 'Un hallazgo avala que el "Hombre de Orce" fue el primer homínido europeo'.

EFE, *El País* 2009, 2 September. 'Las hachas de mano más antiguas de Europa están en Granada y Murcia'.

EFE, *La Vanguardia* 1984, 19 July. 'Se reanuda la investigación sobe el Hombre de Orce': 38.

EFE, *La Vanguardia* 1985, 18 January. 'El descubridor del cráneo de Orce ratifica que el fósil es humano': 21.

EFE, *La Vanguardia* 1996, 27 August. 'Orce reclamará a Sabadell el cráneo hallado por Gibert': 26.

EFE, *Soitu.es* 2008, 6 November. 'Gibert presentará nuevos proyectos en Orce tras el rechazo de 5 propuestas'.

Elgarresta, J. A., *La Verdad Digital* 2001a, 29 January. 'El primer europeo fue cartagenero, asegura el paleontólogo Josep Gibert'.

Elgarresta, J. A., *La Verdad* 2001b, 27 January. 'Cultura pagará la investigación y muestra de los restos hallados en Cueva Victoria'.

El Perich, *El Periódico* 1984, 18 May. 'El Perich'.

Europa Press, *El Ideal de Granada* 2010, 30 August. 'Comienzan este lunes las excavaciones en Orce con el arqueólogo de Atapuera, Robert Sala'.

Europa Press, *El Ideal de Granada* 2013a, 1 March. 'El diente descubierto en Orce (Granada) se convierte en el resto humano más antiguo de Europa'.

Europa Press, *El Ideal de Granada* 2013b, 17 February. 'Las excavaciones en Orce (Granada) se retomarán en agosto'.

Europa Press, *La Vanguardia* 2013, 1 March. 'Descubren en Orce el diente fósil de un homínido, que se convierte en el resto humano más antiguo de Europa'.

Europa Press, *Vilaweb* 2007, 13 October. 'Les cendres de Josep Gibert, descobridor de l'Home d'Orce, reposen en el jaciment de Venta Micena (Granada)'.

Fernández, V., *El Ideal de Granada* 1983, 12 June. 'Importante apoyo a las investigaciones del yacimiento de Venta Micena (Orce)': 17.

Fernández, V., *El Ideal de Granada* 1985, 13 May. 'Granada se retira de la dirección de las excavaciones en Venta Micena (Orce)'.

Fontova, R., *El Periódico* 1990, 20 August. 'Aparecen más restos humanos en Orce': 19.

Fotheringham, A., *The Independent* 2013, 4 March. 'Postcard from ... Orce, Spain'.

Fountain, H. *New York Times* 2009a, 2 September. 'Research Indicates Earlier Debut in Europe For Refined Stone Tools'.

Fountain, H., *New York Times* 2009b, 8 September. 'Earlier Date for First Refined Stone Tools in Europe'.

F.R.A., *ABC* 1995, 9 September. 'La Consejería de Cultura promete apoyo económico para el yacimiento de Orce': 39.

F.R.A., *ABC* 1998a, 20 July. 'El Instituto Crusafont de Sabadell entrega hoy a la Junta los fósiles de Orce': 43.

F.R.A., *ABC* 1998b, 21 July. 'La Junta reitera que mantendrá la paralización de las excavaciones arqueológicas en Orce': 37.

García, A. V., *El País* 1986, 22 August. 'Un estudio científico de la universidad de Granada avala que el "Hombre de Orce" es el homínido más antiguo de Eurasia'.

García-Olivares, E., *ABC* 1996, 20 August. 'Fósiles de Orce': 14.

García-Olivares, E. *El Ideal de Granada*, 2013, 'Orce, 1984'.

Garrido, A., *El Periódico* 1995, 8 September. 'Del Sílex, a Chirac': 7.

Gómez Motero, *ABC* 1983, 10 November. 'Venta Micena puede ser el poblado más antiguo de Europa y Asia': 38.

Gibert, J., *Avui* 1983, 31 August. 'Adéu al mestre Miquel Crusafont': 33.

Gibert, J., *El Ideal de Granada* 1992, 24 January. 'El Hombre de Orce': 18.

Gibert, J., *La Vanguardia* 1993, 1 May. 'El primer ser humano de Europa': 10–11.

Gibert, J., *La Vanguardia* 1996, 4 February. 'Un paleontólogo indignado'.

Gibert, J., *La Vanguardia* 2005, 31 January. 'Asnos y homínidos'.

Gómez, J. E., *El Ideal de Granada* 1997, 3 February. 'La Junta pide a José Gibert que devuelva el Hombre de Orce'.

Gómez, J.E., *El Ideal de Granada* 1997, 20 March. 'Científicos españoles rechazan la humanidad del cráneo de Orce y dicen que es de un caballo'.

Gómez, J. E., *El Ideal de Granada* 1998, 21 July. 'José Gibert pide a la Junta que se reinicien los trabajos en los yacimientos del entorno de Orce': 8.

Gómez, J. E., *El Ideal de Granada* 2007, 16 January. 'El Descubridor de Atapuerca reconoce al Hombre de Orce como el primero que habitó Europa': 9.

Gomis, L., *La Vanguardia* 1984, 16 May. 'El burro de Orce': 27.

González-Besada, J. L., *ABC* 1986, 22 August. 'Silencio sobre el parón en las excavaciones de Venta Micena': 29.

González Casanova, J. A., *El Periódico* 1983, 16 June. 'La nueva generación política': 4.

González, E., *La Vanguardia* 1984, 11 March. 'El Hombre de Orce sigue asombrando'.

González, M. A., *La Vanguardia* 1995, 8 September. 'Piden que Orce sea declarado patrimonio de la humanidad': 42.

Guardia, M. A., *La Vanguardia* 1979, 4 March. 'Lucy: La efímera popularidad de una miss muy mona'.

Guerrero, J., *La Vanguardia* 1984, 13 May. 'La investigación en curso terminará hacia Navidad': 13.

Gurovitz, H., *Folha do São Paolo* 1993, 26 December. 'Arquelogia busca origem de primerios brasilerios'.

Hammond, N., *The Times* 1986, 8 August. 'American visitors 32,000 years ago'.

Hammond, N., *The Times* 1995, 24 January. 'Doubts over date of early American site'.

Hernández, J.J., *Diario de Granada* 1983, 14 June. 'El día más feliz de Orce': 5.

Hernández, B., *El Mundo.es* 2009, 7 September. 'Hallan en Granada las hachas más antiguas de Europa'.

Hurtado, M., *ABC* 1994, 28 July. 'Murcia puede tomar la delantera a Atapuerca': 45.

Hurtado, M., *ABC* 1995, 11 October. 'Expertos afirman que el primer europeo llegó a España a través del Estrecho': 48.

Informativos *Canalsur.es* 2013, 2 March. 'Gibert: "Me duele que desacrediten a mi padre, pero me alegra lo del diente" '.

J.A., *Diari de Sabadell* 1983a, 29 June. 'El "Hombre de Venta Micena", huésped de honor'.

J.A., *Diari de Sabadell* 1983b, 17 June. 'Sabadell, meca mundial de la Paleontología': 12–13.

J.A., *Diari de Sabadell* 1984, 28 January. 'Es muy probable la aparición de nuevos hombres de Orce': 12–13

J.A., *Diari de Sabadell* 1986a, 29 November. 'Tots són deutors i hereus del Dr. Crusafont': 3.

J.A., *Diari de Sabadell* 1986b, 29 November. 'Expectación y escepticismo en la presentación del Hombre de Orce'.

J.A., *Diari de Sabadell* 1986c, 30 August. 'Josep Gibert: "Voy a convencer a quienes todavía duden de mi" ': 1, 18–19.

J.A., *Diari de Sabadell* 1990, 25 August. 'El Dr. Josep Gibert troba més homes a Orce': 7.

J.J.N.A., *El País* 1984, 18 May. 'Josep Gibert: "Quiero mantenerme en el terreno científico" '.

J.J. Navarro Arisa, *El País* 1984, 15 May. 'El director del Museo del Hombre de París no descarta que el fósil de Orce sea humano'.

J.M., *Granada Hoy* 2005, 13 March. 'El investigador acudirá a la justicia si no puede trabajar en Venta Micena': 3.

Jover, J., *La Vanguardia* 1985, 11 October. 'La Guardia Civil desaloja de Cueva Victoria a 7 arqueólogos catalanes': 35.

Pérez Piñar, J. *El Periódico* 1985, 10 February. 'Un paseo insólito por el túnel de los tiempos': 24.

Keys, D., *The Independent* 1995, 8 October. 'First European' lived in Spain'.

Ladrón de Guevara, J. G., *El Ideal de Granada* 2007, 22 January. 'Muy buenas noticias'.

Leite, M., *The Guardian* 1991, 20 December. 'A war of scientific worlds in the Brazilian hinterland': 28.

L.L.C. *ABC* 2015, 24 March. 'La "pinza" acabó en veinte meses con el único gobierno en minoría en Andalucía'.

Lopes, R. J., *Folha de São Paolo* 2015, 12 February. 'Piauí teria ocupação humana há 20 mil anos'.

Lyon, W., *El País* 1983a, 15 June. 'Los hijos de Adán eran andaluces'.

Lyon, W., *El País* 1983b, 19 June. 'Los dos hombres de Orce'.

MacPherson, A., *El Periódico* 1986, 19 August. 'El Hombre de Orce sacaba el tuétano de los animales': 10.

Madridejos, A., *El Periódico* 2009, 2 September. 'Las hachas más antiguas de Europa "aparecen" en Murcia'.

Madridejos, A. *El Periódico* 2013, 1 March. 'Descubierto en Orce el diente fósil de un homínido de hace 1,4 millones de años'.

Manso, S., *ABC* 1984a, 23 February. 'Se confirma la antigüedad del hombre de Orce : un millón quinientos mil años': 47.

Marot, J., *der Standard* 2013, 8 May. 'Wirbel um die Wiege der Europäer': 17.

Martínez, I., *ABC* 2001, 6 August. 'Tras las huellas de nuestros antepasados': 45.

Mercade, J., *La Vanguardia* 1986a, 28 November. 'La Sociedad de Paleontología homenajeó a Miquel Crusafont': 41.

Mercade, A. *La Vanguardia* 1986b, 29 November. 'Una ponencia de Gibert cerró las Jornadas de Paleontologia': 41.

Mercadé, J., *La Vanguardia* 1987, 2 October. 'Los colaboradores de Gibert dicen que el cráneo de Orce pertenece a un equino': 40.

Moix, L., *La Vanguardia* 1985, 2 February. 'Josep Gibert confirma que la falange hallada en Murcia es el resto humano más antiguó de Europa': 25.

Montero, G., *ABC* 1983, 6 June. 'Granada, posible cuna del hombre euro-asiático': 36.

Morales, M., *El Periódico* 1984, 16 May. 'martinmorales'.

Morán, C., *El Ideal de Granada* 1995, 5 September. 'La comunidad científica respalda la tesis de que los restos humanos de Orce son los más antiguos de Europa': 4.

Moreno, M., *El País* 1983, 15 June. 'El descubrimiento del "Hombre de Orce" trastoca las actuales teorías sobre la vida humana en el continente euroasiático'.

Moreno, M., *El País* 1986, 9 September. 'El cráneo de Orce es humano, según confirma el Congreso Arqueológico Mundial'.

Moreno, M., *El País* 1990, 11 February. 'Josep Gibert asegura haber descubierto nuevos restos del hombre de Orce'.

Moyà-Solà, S.; Köhler, M. *La Vanguardia* 1996, 4 February. 'Paleontólogos agradecidos'.

Moyà-Solà, S., *El País* 2007, 10 October. 'Josep Gibert, paleontólogo'.

M. S., *ABC* 1983, 6 July. 'Los restos del "Hombre de Orce", en una caja fuerte': 42.

M. S., *Vilaweb* 2008, 6 February. 'Xavier Folch, un homenatge'.

Muñoz, J., *Granada Hoy* 2005, 13 March. 'La Junta multa a Gibert por excavar sin permiso en Orce': 2.

Natale, D., *Folha de São Paolo* 1984, 10 September. 'Arqueólogos alertam para aventuras de amadores'.

Neves, W., *Folha do São Paolo* 1995, 8 May. 'Pôlémica de Pedra Furada'.

Oliveras, E., *El Periódico* 1983, 12 November. 'Una feria rodante muestra Andalucia a toda España': 20.

Ortuño, J. I., *Diari de Terrassa* 1987, 12 February. 'Investigando un "crimen" de hace un millón trescientos mil años'.

Otr., *ABC* 1984, 18 July. 'El profesor Guibert [sic] excava de nuevo en el yacimiento de Orce': 29.

Our Madrid Correspondent, *Daily Telegraph* 1984, 14 May. 'Ass is Taken for a Man': 16.

Palmqvist, P. *El Ideal de Granada* 1996, 15 November. 'Carencia de Método Científico'.

Paredes, J. A.; Moix, L. *La Vanguardia* 1984, 26 January. 'Venta Micena es el asentamiento humano más antiguo de Eurasia': 26.

Paredes, J. A., *La Vanguardia* 1986, 23 August. 'Gibert: las dudas sobre el "Hombre de Orce" acabarán en Southampton'.

Pedregosa, F., *Diario de Granada* 1983, 12 June. 'El Homínido más antiguo de los descubiertos en Eurasia': 3.

Pedregosa; F., *El Periódico* 1983, 26 June. 'Los paisanos del hombre más viejo'.

Pernau, J., *El Periódico* 1983, 16 June. 'Todos somos andaluces': 5.

Pernau, J., *El Periódico* 1984, 13 May. 'Descendemos del Burro': 7.

Piñol, R. M., *La Vanguardia* 1984, 15 March. 'La nova editorial catalana Empúries publicará diverses obres de Riba': 43.

Piñol, R. M., *La Vanguardia* 1986, 3 September. 'Gibert no fue discutido en Southhampton al presentar las pruebas sobre el cráneo de Orce': 28.

Planas, M., *La Vanguardia* 2007, 9 October. 'Un hombre marcado por un fòsil': 37.

Prats, C., *La Vanguardia* 1984, 18 April. 'Castellón: hallan un asentamiento humano de dos millones de años': 33.

Quero, L., *El País* 2002, 6 July. 'El principal yacimiento de Orce queda otra vez al margen de las excavaciones'.

Quero, L., *El País* 2003, 4 April. 'La Junta autoriza nuevas excavaciones en el principal yacimiento de Orce'.

Ramentol, S., *El Periódico* 1983a, 14 August. 'La Asombrosa Aventura del Hombre de Orce'.

Ramentol, S., *El Periódico* 1983b, 24 June. 'Peligran las excavaciones en Orce por falta de medios': 15.

Ramos Espejo, A., *Diario de Granada* 1983, 19 June. 'Hijos del Homínido': 10.

Rebeyrol, Y., *Le Monde* 1986, 9 July. 'Les ancêtres de Nouveau Monde'.

Redacción, *ABC* 1983a, 24 June. 'La antigüedad atribuida al hombre de Orce es excesiva': 16.

Redacción, *ABC* 1983b, 27 June. 'El Hombre de Orce, trasladado a Sabadell': 38.

Redacción, *ABC* 1983c, 29 June. 'Polémica en Granada por el traslado a Sabadell del Hombre de Orce': 36.

Redacción, *ABC* 1983d, 19 July. 'Los restos del homínido de Orce volverán a Andalucía': 12.

Redacción, *ABC* 1983e, 9 July. 'La hora "H" ': 87.

Redacción, *ABC* 1983f, 19 June. 'Ya tiene Consejo el Departamento de Estudios Andaluces': 48.

Redacción, *ABC* 1983g, 22 October. 'Presentados dos nuevos cuadernos de Historia de andalucía': 40.

Redacción, *ABC* 1995a, 6 September. 'El primer europeo': 16.

Redacción, *ABC* 1995b, 10 June. 'El Hombre de Orce, primer europeo': 55.

Redacción, *ABC* 1996c, 11 August. 'Polémica en Orce': 49.

Redacción, *ABC* 2001, 24 July. 'Los restos del elefante refuerzan la hipótesis de que Orce tuvo presencia humana': 8.

Redacción, *ABC* 2002, 24 August. 'La Junta permite continuar los trabajos en el yacimiento de Orce': 51.

Redacción, *ABC* 2003, 2 August. 'Comienza en Orce el estudio del resto faunístico más importante de Europa': 55.

Redacción, *El Defensor de Granada* 1983, 22 July. 'Granada Discriminada': 1.

Redacción, *El Ideal de Granada* 2009, 12 November. 'Un arqueólogo de Atapuerca dirigirá excavaciones en Orce'.

Redacción, *El Ideal de Granada* 2013a, 4 March. 'IU propone a José Gibert como hijo adoptivo de la Provincia de Granada'.

Redacción, *El Ideal de Granada* 2013b, 6 November. 'El pintor José Guerrero será hijo predilecto de la provincia de Granada'.

Redacción, *El Mundo* 1996, 11 January. 'Jordi, el eslabón perdido'.

Redacción, *El Mundo* 1999, 26 February. 'El hombre de Orce llegó a Europa a través del Estrecho Cruzó hace 2,4 millones de años, al bajar el nivel del mar y aparecer un islote situado a cinco kilómetros de Africa y tres de la Península, según el director del yacimiento'.

Redacción, *El Mundo* 2004, 16 September. 'Ha estado con nosotros … José Maria Bermúdez de Castro'.

Redacción, *El Mundo* 2005, 14 October. 'Ha estado con nosotros … Juan Luis Arsuaga'.

Redacción, *El País* 1984a, 12 May. 'Un gran descubrimiento presentado sin reservas': 23.

Redacción, *El País* 1984b, 15 May. 'Hombre o borrico'.

Redacción, *El País* 1984c, 13 May. 'Motivos del Aplazamiento'.

Redacción, *El País* 1984d, 12 May. 'Un homínido con poca barbilla': 23.

Redacción, *El País* 1985, 2 March. 'Críticas a la "imprudente" difusión de los hallazgos de Gibert'.

Redacción, *El País* 1987, 2 October. 'Dos investigadores del cráneo de Orce afirman que perteneció a un équido joven'.

Redacción, *El Periódico* 1983a, 17 July. 'El Hombre de Orce vuelve esta semana a Granada'.

Redacción, *El Periódico* 1983b, 13 December. 'Hoy'.

Redacción, *El Periódico* 1983c, 13 July. 'Hoy': 35.

Redacción, *El Periódico* 1983d, 4 November. 'Expertos de todo el mundo tratarán del Hombre de Orce'.

Redacción, *El Periódico* 1984a, 11 January. 'Hombre de Orce': 1.

Redacción, *El Periódico* 1984b, 15 May. 'Encuentro sobre Prehistoria en Perpiñán'.

Redacción, *El Periódico* 1985a, 28 February. 'Gibert sale en defensa de su Hombre de Orce': 23.

Redacción, *El Periódico* 1985b, 20 March. 'El fósil de Orce no podrá revelar su carácter humano': 19.

Redacción, *El Periódico* 1986a, 18 May. 'Al final, los prosoviéticos se presentan en solitario'.

Redacción, *El Periódico* 1986b, 3 September. 'Presentan las pruebas del orígen humano del cráneo de Orce': 17.

Redacción, *El Periódico* 1986c, 9 September. 'Ratificado el orígen humano del cráneo de Orce': 16.

Redacción, *El Periódico* 1986d, 30 November. 'Ponen en duda el orígen del Hombre de Orce': 28.

Redacción, *Folha do São Paolo* 1978, 30 June. 'Arte Rupestre': 32.

Redacción, *Folha do São Paolo* 1979, 23 April. 'Piauí, por aqui': 25.

Redacción, *Folha do São Paolo* 1993, 11 April. 'Teoria tradicional é falha, diz Guidon'.

Redacción, *Granada Digital* 2005, 11 August. 'Los colmillos de elefante hallados en Orce servirán para analizar las condiciones climáticas de hace 1.500.000 años'.

Redacción, *Granada Hoy* 2005, 13 March. 'Gibert, multado por excavar en Orce sin permiso, amenaza con denunciar a la Junta': 1.

Redacción, *Granada Hoy* 2009, 5 April. 'Los que deciden las excavaciones de Orce son mucho más cualificados que Gibert'.

Redacción, *La Vanguardia* 1982, 1 July. 'Libros: Recién Salidos': 38.

Redacción, *La Vanguardia* 1983a, 5 June. 'Importante hallazgo paleontológico en Granada': 38.

Redacción, *La Vanguardia* 1983b, 26 June. 'El "Hombre de Orce" será estudiado en el centro de Paleontología de Sabadell': 35.

Redacción, *La Vanguardia* 1983c, 17 August. 'El paleontólogo Crusafont ha fallecido': 17.

Redacción, *La Vanguardia* 1984a, 2 February. 'Día a Día': 15.

Redacción, *La Vanguardia* 1984b, 22 February. 'Dia a Dia': 53.

Redacción, *La Vanguardia* 1984c, 13 May. 'Ningún elemento actual permite atribuir el cráneo de Orce a un asno': 1.

Redacción, *La Vanguardia* 1988, 18 February. 'Interesante Conferencia a cargo del Dr. Josep Gibert en la Casa de Madrid, Barcelona': 40.

Redacción, *La Vanguardia* 1989, 11 April. 'Conferencias': 87.

Redacción, *La Vanguardia* 1990, 20 August. 'Hallados nuevos restos del Hombre de Orce': 23.

Redacción, *La Vanguardia* 1995, 5 September. 'Un congreso internacional asegura que el hombre de Orce es el europeo más antiguo': 36.

Redacción, *La Vanguardia* 1997, 18 August. 'Los científicos hallan pruebas demoledoras contra la autenticidad del hombre de Orce': 20.

Redacción, *Mediterráneo* 1983, 15 June. 'Descubrimiento de homínido en Granada': 23.

Redacción, *Mediterráneo* 1984, 18 April. 'Restos de actividad humana con dos millones de años de antigüedad': 5.

Redacción, *Noticiero Universal* 1983, 29 June. 'Subvención oficial para las excavaciones en Orce'.

Redacción, *RadioGranada.es* 2013, 1 March. 'Sí hay Hombre de Orce: El diente fósil de un homínido consigue el aval de la comunidad internacional como el resto humano más antiguo de Europa'.

Reis, J., *Folha do São Paolo* 1986, 18 October. 'Homem pode ter vivido no Piauí há 32 mil anos'.

Relaño, A., *El País* 1984a, 12 May. 'Serios indicios de que el cráneo del 'hombre de Orce' pertenece a un asno': 1.

Relaño, A., *El País* 1984b, 12 May. 'Serias sospechas de que el 'hombre de Orce' no es un hombre, sino que podría ser un asno de dos a cuatro meses de edad': 22.

Relaño, A. *El País* 1984c, 13 May. 'La Junta de Andalucía valora con cautela las dudas sobre el origen del Hombre de Orce'.

Relaño, A. *El País* 1984d, 14 May. 'El pueblo granadino de Orce, indiferente ante la polémica sobre el origen del cráneo encontrado en Venta Micena'.

R.F. *El Noticiero Universal* 1983, 13 December. 'El Hombre de Orce atrae a la paleontologia mundial': 14.

Ribot, F. *ABC* 1996, 27 August. 'Cráneo de Orce': 14.

Ribot, F. *El Ideal de Granada* 2007, 20 February. 'El Cráneo de Orce'.

Rico, B. *Granada Hoy* 2013, 26 March. 'Human Evolution pide a los autores corregir el artículo del diente de Orce'.

Rivera, A. *El País* 1995, 6 October. 'Paleontólogos y geólogos cuestionan la existencia del Hombre de Orce': 30.

Rodríguez, G. *La Opinión de Granada* 2007, 22 January. 'El Galileo de Orce': 12.

Román, F. J. *El País* 1998, 1 August. 'JOSÉ LUIS ARSUAGA ARQUEÓLOGO: "Los restos de Orce no son humanos, no tengo ninguna duda" '.

Romero, R., *Granada Hoy* 2007, 16 January. 'Gibert reabre con nuevas pruebas el debate sobre el hombre de Orce'.

Romero, S., *The New York Times* 2014, March 28. 'Discoveries Challenge Beliefs on Humans Arrival in the Americas'.

Ruiz Antón, F. *ABC* 1994, 19 October. 'IU exige a la consejería de Cultura mayor apoyo al yacimiento arqueológico de Orce': 42.

Ruiz Antón, F. *ABC* 1995a, 6 September. 'El hombre de Orce vivía en una playa tranquila al borde de un lago y se disputaba la carne con las hienas': 39.

Ruiz Antón, F. *ABC* 1995b, 8 September. 'Los científicos piden que los yacimientos de Orce sean declarados Patrimonio de la Humanidad': 38.

Ruiz Antón, F. *ABC* 1995c, 4 September. 'Más de doscientos especialistas participan desde hoy en Orce en el Congreso de Paleontología Humana': 31.

Ruiz Antón, F. *ABC* 1999, 12 July. 'Orce: un nuevo equipo de científicos reanuda hoy las excavaciones': 39.

Ruiz de Elvira, M. *El País* 1997, 17 August. 'El fósil del Hombre de Orce pertenece en realidad a un caballo, según los científicos'.

S.A. *El Faro de Cartagena* 2001, 26 January. 'Josep Gibert data los fósiles de Cueva Victoria como los más antiguos de Europa': 12.

Sáiz-Pardo, M. *El Ideal de Granada* 2007, 16 January. 'Orce, como Atapuerca': 5.

Sanz, R. *ABC* 2007, 9 October. 'Y el Hombre de Orce saltó a la Palestra': 59.

S.C., *ABC* 1994, 31 July. 'Si el "primer europeo" levantase la cabeza ...': 49.

S.C., *ABC* 1995, 5 September. 'Los paleontólogos sentencian que el Hombre de Orce fue el primer europeo': 49.

S.C., *ABC* 1997, 21 August. 'El Ayuntamiento de Orce reclama los permisos para seguir excavando': 42.

S.C., *ABC* 1998, 25 March. 'Dos revistas científicas publican nuevos trabajos que avalan la antigüedad del Hombre de Orce': 55.

Sella, A., *La Vanguardia* 1996, 3 February. 'Paleontólogos a la Greña' *Ciencia y Vida*:10–13.

Soria, A., *ABC* 2004, 18 June. 'Josep Gibert defiende la autenticidad de su Hombre de Orce en su nuevo libro': 58.

Sotos, E.; Madridejos, A. *El Periódico* 2015, 2 March. 'Un yacimiento en Brasil reescribe el primer poblamiento de América'.

S.S., *ABC* 2009, 3 September. 'Descubren en Murcia y Granada las hachas más antiguas de Europa'.

S.S.; J.M., *Granada Hoy* 2005, 2 July. 'Los equipos vuelven a Orce con las mejores expectativas en años'.

Strauss, S. *The Globe and Mail* 1987, 17 October. 'Did humans roam Americas millenia before Ice Age?'.

Teresina (Correspondente), *Folha de São Paolo* 1978, 10 February. 'Arqueólogos farão pesquisas no Piauí': 19.

Titos, F. J., *El País* 1999, 4 January. 'Pascual Rivas, Paleontólogo en Orce: "No se puede atraer gente a un museo para ver un fósil dudoso" '.

Trillo, M.; Muñoz, J., *Granada Hoy* 2007, 2 March. '¿Qué pasó en Orce el verano de 2003?': 4.

Troyano, R., *ABC* 1995, 5 September. 'Expertos mundiales confirman en Orce la autenticidad de los restos humanos encontrados, los más antiguos de Eurasia': 41.

Troyano, R., *ABC* 1997, 9 December. 'La Junta no concederá nuevos permisos de excavaciones en Orce, según el director de Bienes Culturales': 50.

Troyano, R., *ABC*, 1998a, 9 March. 'El alcalde de Orce pide a la Junta permiso para nuevas excavaciones arqueológicas': 44.

Troyano, R., *ABC* 1998b, 13 October. 'El Ayuntamiento mejora el museo donde se expondrá el Hombre de Orce': 52.

Troyano, R., *ABC* 1999, 28 March. 'Las excavaciones en Orce se reiniciarán en un par de semanas': 43.

Turbón, D., *La Vanguardia* 1995, 17 September. 'Hombre de Orce: Josep Gibert Tenía razón': 64.

Utrera, J. *El Ideal de Granada* 2005a, 19 January. 'La Generalitat catalana distingue al profesor Gibert con la medalla al mérito científico'.

Utrera, J. *El Ideal de Granada* 2005b, 5 July. 'Cultura retoma las investigaciones en el yacimiento de Venta Micena'.

Utrera, J. *El Ideal de Granada* 2009, 4 September ' "Nature" reconoce la presencia de restos humanos en Orce'.

Utrera, J. *El Ideal de Granada* 2010, 14 September. 'La huella humana en Orce es la más antigua de Europa'.

Utrera, J. *El Ideal de Granada* 2012, 26 June. 'Cultura deja desierto el concurso para excavar en Orce'.

Utrera, J. *El Ideal de Granada* 2013, 24 March. 'La revista Journal of Human Evolution retira temporalmente el artículo sobre el diente humano de Orce'.

Utrera, J. *El Ideal de Granada* 2015, 6 August. 'Inaugurado en Orce el Centro de los Primeros Pobladores Europeos'.

Valenzuela, A., *ABC* 2006, 2 April. 'Es imposible que el primer europeo se encuentre en Atapuerca': 29.

Valverde, F., *El País* 2010, 12 September. 'Los mamuts del hombre de Orce'.

Vilardebó, I., *ABC* 1984, 1 September. 'La importancia arqueológica del yacimiento de Orce es hoy incuestionable' *Sábado Cultural*: IV–V.

Vilardebó, I., *ABC* 1985, 19 May. 'Granada : Retirada andaluza en la dirección de las excavaciones de Orce': 54.

Vilardebó, I., *ABC* 1986a, 20 August. 'En el Congreso Mundial de Arqueología se defenderá el carácter homínido del cráneo de Orce': 25.

Vilardebó, I., *ABC* 1986b, 20 August. 'Los descubridores del Hombre de Orce buscan respaldo internacional': 37.

Vincent, I. *The Globe and Mail* 1994, 1 January. 'Research in Brazil suggests early man arrived from several directions up to 50.000 years ago'.

Wigg, R., *The Times* 1983, 16 July. 'Scientists in search of Europe's earliest man': 6.

Weimer, W. *Frankfurter Allgemeine Zeitung* 1995, 7 September. 'Wo lebte der erste Europäer?'.

Weise, E., *USA TODAY* 2009, 2 September. 'Europe gains an edge in ax mystery'.

Wilford, J., *The New York Times* 1986, 19 June. 'Far earlier evidence of humans in Americas'.

Wilford, J., *Journal do Brasil* 1986 20 June. 'Homem ocupou América há 32 mil anos'.

Wilford, J., *The New York Times* 1995, 14 February. 'Doubts Cast on Report Of Earliest Americans'.

W.L., *El País* 1983, 15 June. 'Huesos extaños'.

Wirth, R. *La Vanguardia* 1999, 3 November. 'Daniel Turbón, entrevista' *Vivir en Barcelona*: 2.

Index

www.ingramcontent.com/pod-product-compliance
Lightning Source LLC
Chambersburg PA
CBHW071323230926
51849CB00002B/303